Digital Design for Computer Data Acquisition

Digital Design for Computer Data Acquisition

Charles D. Spencer

Department of Physics
Ithaca College

CAMBRIDGE UNIVERSITY PRESS

CAMBRIDGE
NEW YORK   PORT CHESTER   MELBOURNE   SYDNEY

Published by the Press Syndicate of the University of Cambridge
The Pitt Building, Trumpington Street, Cambridge CB2 1RP
40 West 20th Street, New York, NY 10011, USA
10 Stamford Road, Oakleigh, Melbourne 3166, Australia

First published 1990

Printed in the United States of America

Library of Congress Cataloging-in-Publication Data
Spencer, Charles D.
Digital design for computer data acquisition / Charles D. Spencer
p. cm.
Includes bibliographical references.
ISBN 0–521–37199–6
1. Digital electronics. 2. Computer interfaces. 3. Automatic
data collection systems. I. Title
TK7868.D5S65 1990
621.39'81–dc20   7/969   89–71224

British Library Cataloguing in Publication Data
Spencer, Charles D.
Digital design for computer data acquisition.
1. Electronic equipment. Digital circuits. Design
I. Title
621.3815
ISBN 0–521–37199–6

Analog Devices® is a registered trademark of Analog Devices, Inc. Apple® and Applesoft BASIC® are registered trademarks of Apple Computer, Inc. Macintosh™ is a trademark licensed to Apple Computer, Inc. IBM PC®, XT™, AT®, and PS/2® are registered trademarks of International Business Machines, Inc. Intel™ is a trademark of Intel Corp. Microsoft® and QuickBASIC™ are registered trademarks of Microsoft Corp.

To Eulalah, Bob, and Susie

Contents

Preface

Advances in electronics during the past decade have led to a variety of computer based instruments. Manufacturers continually add flexibility to traditional research tools as they introduce devices not previously feasible. With measurements made by computers, keyboards replace switches, real-time graphics is common, powerful software is ready to analyze results, and data are easily saved and/or transferred to larger computers. Another important development is adaptability. Minor changes in software and/or hardware can significantly alter a system's capabilities. This book is for people interested in both digital design and adaptable computer based data acquisition.

The Parallel Data Collector (or PDC) has been developed at Ithaca College. It works on Apple II and on IBM PC/XT/AT and compatible computers. With commercially available I/O ports, it also works with Macintosh and IBM PS/2 machines. The PDC can be a frequency meter, voltage recorder, pulse height analyzer, multichannel scaler and many other instruments. The system supports voltage, time and counting measurements with programmable parameters such as voltage conversion rate, time accuracy and counting interval. This book first describes the design principles and integrated circuits which comprise the PDC and then explains the system itself. The central idea is that it's better to master and apply a few concepts than to acquire a broad background with no particular objective in mind.

This book is mostly a text on digital electronics and interfacing. But it also tells "how to" build a powerful, adaptable data acquisition system. Problems at the end of each chapter reinforce the material and many are suitable for laboratory exercises. The later chapters present the PDC's hardware and software as well as construction details and guidelines for testing and troubleshooting. Every circuit has been built and used and every program entered and evaluated. No previous knowledge of digital electronics is assumed.

Chapter 1 outlines how computers communicate with outside circuitry and reviews commercial data acquisition systems.

Chapter 2 introduces digital electronics starting with basic logic gates and ending with tri-state bus drivers.

Chapter 3 describes parallel I/O ports which link a host computer (and its software) with data acquisition circuitry. Designs are explained for the expansion slots of Apple II's and IBM PC/XT/AT and compatibles. Construction and testing details are in Appendices D and E. Vendors are listed for commercial ports for these systems and for IBM PS/2's and Macintoshes. Also, the three software concepts needed for PDC operations are explained and illustrated.

Chapters 4-6 are tutorials on three specific devices: the AD573 Analog-to-Digital Converter (ADC), the 6264 static memory and the 8253 Programmable Interval Timer (PIT).

Chapter 4 presents the 10-bit, 20 μsec AD573. After covering basic operations, circuitry which interfaces the device to the parallel ports is presented. Control software is illustrated.

Chapter 5 explains the operation of the 6264 8K by 8-bit static RAM. Interface circuitry is presented along with a program which loads values in and afterwards retrieves values from the memory. Hardware and software for a 1.0000 MHz digital recorder are given.

Chapter 6 presents the Intel 8253 PIT. The focus is on control signal generation and counting measurement. After describing how to interface the chip to the parallel ports, software for all operations is given. To illustrate the use of PIT's in instruments, a programmable scaler is explained. Later, PIT's become the most important PDC building block.

Chapters 7 and 8 cover the PDC's control and ADC modules. Guidelines for construction, testing and troubleshooting are given.

Chapter 9 shows how to make the PDC into a comprehensive measurement system with the capability to repeatedly determine two voltages, count events and measure time, all in a variety of formats with multiple programmable parameters. Numerous examples are given along with details on construction, testing and troubleshooting.

Chapter 10 explains the hardware and software for a programmable rate "fast" voltage recorder. Construction and testing details are included.

All circuit diagrams and program listings in this book have been double-checked for errors. However, neither I nor the publisher warranties the information. I have made every effort to document all trademarks.

I wish to thank Professor Jack Peck of the Ithaca College Psychology Department for reading the manuscript and making many worthwhile suggestions. Erik Herrmann checked the manuscript for technical errors. I'm indebted to my physics colleagues, Professors Peter Seligmann and Dan Briotta, for many of the ideas incorporated in the PDC. Also, the following Ithaca College physics and physics-computing majors built and tested PDC hardware and software: Doug Wilson, Mike Dailey, Matt Dubois, Jerry Bush, Brendan Madden, Jerry Walker, J. B. Chupick and Erik Herrmann.

Charles D. Spencer
Ithaca, New York

1 Computer Measurement

What does someone do who wants to use computers for data acquisition? No matter which way the goal is pursued, it's necessary to know how computers communicate with the external circuits which make measurements. This chapter summarizes choices and introduces the approach used in this book.

1.1 Ports and Expansion Slots

Measurement Circuits (MC's) can be connected to computers through game ports, serial ports and expansion slots. In each case, software must send values to and get values from the MC's. The following sections review and compare the approaches.

1.1.1 Game Port Communication

On an Apple II, communication between its game port and an MC occurs as follows. Up to three digital signals generated by an MC are connected to the three game port push buttons. The computer ascertains a signal's high or low state by peeking a specified address and then seeing if the value is greater than 127 which means the signal is high. Similarly, one or more of the game port's four enunciator outputs are connected to an MC. The computer makes an output high by peeking a specified address and low by peeking a different address.

Game port data acquisition is practical when the number of input and output bits can be limited which is the case with the few commercially available systems. They are typically self-contained units starting with, for instance, a temperature probe and ending with a data plot. Their low cost, reliability and operational simplicity make them attractive for instructional laboratories.

1.1.2 Serial Communication

The most widely adopted standard in computers is RS-232 serial communication. For virtually any machine, serial ports either are or can be installed (at low cost).

To communicate serially, an MC must have a Universal Asynchronous Receiver/Transmitter (UART). This large scale integrated circuit receives a byte one bit at a time from the computer's serial port and makes the value available as eight separate signals. Similarly, the chip takes eight signals (generated by the MC) and sends them one bit at a time to the computer's serial port which assembles the value. To realistically set up and control the UART, the MC needs its own microprocessor, memory and software. This means serial data acquisition systems are also computers.

Even at maximum bits-per-second, serial communication is relatively slow. Therefore, measuring one voltage thousands of times a second or several voltages hundreds of times a second requires the MC to temporarily store data values in its own memory. Then, after acquisition, the host computer inputs the values. Data acquisition and storage are most efficiently carried out by the same microprocessor which manages the UART.

Serial systems are necessarily complex and require two levels of software: programs for the MC's processor and programs for the host computer (which include calls to the system software which operates the serial port).

A number of excellent serial data acquisition systems are available for popular computers. They are typically self-contained and come with all hardware and software.

1.1.3 Parallel Communication

Apple II, Macintosh II, IBM PC/XT/AT, IBM PS/2 and all compatibles have expansion slots. Communication between an MC occupying a slot and the computer is fast and easy. For systems with Intel microprocessors, an 8- or 16-bit value is sent to an MC by execution of an out to I/O port. Similarly, an 8- or 16-bit value is obtained from an MC by execution of an input from I/O port. With 8 or 16 bits sent or received in parallel, communication is hundreds of times faster than is possible with serial or game ports.

Data acquisition systems with every imaginable capability are available for the expansion slots of popular computers, especially IBM PC's.

Since these systems work only in the expansion slot for which they were designed, they cannot be moved from say an IBM PC to an IBM PS/2.

1.2 GPIB Standard

Because serial MC's are overly complex and expansion slot systems are not portable, a need exists for a fast parallel interface which makes different computers "look" the same to data acquisition circuitry. Of the several standards proposed over the years, by far the most successful is the IEEE-488 General Purpose Interface Bus (GPIB). It presents to an MC eight connections for sending and receiving data and eight management lines. The protocol is sufficiently elaborate that multiple MC's can be simultaneously interfaced. In addition, a variety of printers, plotters and other devices work through GPIB.

GPIB systems may be purchased for any computer with expansion slots. Versions are also available which interface serially (although their operation is slower). GPIB's usually come with software which initializes and sets up the system and then communicates with whatever MC's and other devices are connected.

Because GPIB makes different computers "look" the same, manufacturers can produce a variety of MC's knowing the market is independent of whatever computer is currently popular. And users can purchase MC's knowing they can upgrade computers without replacing their data acquisition systems.

1.3 Parallel I/O Ports

Another communications approach has many of GPIB's advantages and some of its own. Parallel I/O ports can be acquired for any computer with expansion slots. An OUT (or POKE) operation sends an 8- or 16-bit value to a port where it remains until the next OUT (or POKE). An IN (or PEEK) operation gets the 8- or 16-bit value currently at an input port. MC's connected to the ports supply the inputs and use the outputs.

The ports make different computers "look" the same to MC's and have the advantage of fast parallel communication, the same as GPIB. However, the ports require simpler and less expensive hardware. And software is just sequences of OUT's and IN's (or POKE's and PEEK's) as opposed to calls to system routines supplied by the manufacturer.

Parallel ports are a reasonable interface for data acquisition systems. Also, they are especially helpful for the teaching goals of this book. Their design in Chapter 3 illustrates and reinforces the digital electronics in Chapter 2. After visualizing IN's and OUT's, Chapters 4, 5 and 6 explain and demonstrate how software uses the ports to operate an analog-to-digital converter, an 8K static RAM and a programmable interval timer. The later chapters show how to combine these devices into a variety of useful MC's.

2 Introduction to Digital Electronics

Modern Integrated Circuits (IC's) have changed the effort and skills required to design and construct systems which carry out sophisticated digital operations. While it's desirable to understand how IC's work in terms of semiconductor physics and while good engineering practices are necessary in commercial products, it's now possible to successfully put together data acquisition hardware without first mastering these subjects.

Only fifteen different IC's are required for the parallel ports circuits of Chapter 3, for the analog-to-digital converter, memory and programmable interval counter circuits of Chapters 4-6, and for the measurement systems of Chapters 7-10. The purpose of this chapter is to introduce these and a few other IC's as well as associated design principles. Section 2.1 covers basic logic gates. Section 2.2 introduces integrated circuits, breadboarding and troubleshooting. Section 2.3 presents digital clocks. Section 2.4 introduces the 7474 flip-flop and a variety of timing and control operations. Section 2.5 outlines electrical properties of logic gates. Section 2.6 presents binary, hexadecimal and decimal number systems. Section 2.7 covers the 74192 and 74193 counters and several applications. Section 2.8 presents the 74138 decoder/demultiplexer and additional control applications. Section 2.9 introduces the 74152 multiplexer. And Section 2.10 presents tri-state gates, buses and the 74373 and 74244 buffers. While these topics provide a complete background for this book, readers are referred to Appendix A for more comprehensive introductions to digital electronics.

In general, a logic gate may be thought of as shown in Figure 2.1. It has one or more inputs and one or more outputs, and requires +5 volts and ground. The inputs must be in one of two states (+5 volts or zero volts). The outputs, with an exception discussed later, must be in one of the same two states. Depending on the circumstances, +5 volt inputs and outputs are thought of as **true, on, high** or the digit **1**. Zero volt inputs and outputs are referred to as **false, off, low** or the digit **0**.

Actually, low is indicated by a range of voltages around 0 and high by a larger range around +5. Sometimes it's necessary to worry about where low ends and high begins, but not here.

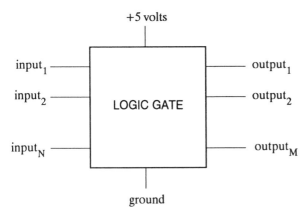

Figure 2.1. Logic gates have one or more inputs and one or more outputs. Inputs and outputs are either high (+5 volts) or low (zero volts).

Logic gates are combined to carry out desired operations. This is accomplished on two levels: symbolic and actual. Circuits are thought about in efficient symbolic terms. Afterward, they are implemented by wiring together actual integrated circuits. The approach in this chapter is to first present logic operations and then discuss building, testing and troubleshooting procedures.

2.1 Basic Logic Gates

Basic logic gates are characterized by one or more inputs and a single output. A common way to specify a gate's operation is with a truth table which gives the output for every possible combination of inputs.

Only three basic gates are required for all circuits in this book: AND, OR, and INVERTER. Their circuit symbols, logic operations and truth tables are given in Figure 2.2. For completeness, NAND, NOR and EXCLUSIVE-OR are described. For all gates except the INVERTER, it's straightforward to extend the logic operation, circuit symbol and truth table from two to three or more inputs.

As can be seen from the truth table, the output of an AND gate is high only if all inputs are high. On the other hand, the output of an OR gate is high if only one input is high. An INVERTER gate's output is always opposite the input. NAND and NOR gates are the same as AND and OR, respectively, with the outputs inverted. Finally, the output of an EXCL-OR is high if two inputs are not equal and low only when all inputs are equal. The following sections show how to use the truth tables in Figure 2.2 to think about and apply basic gates.

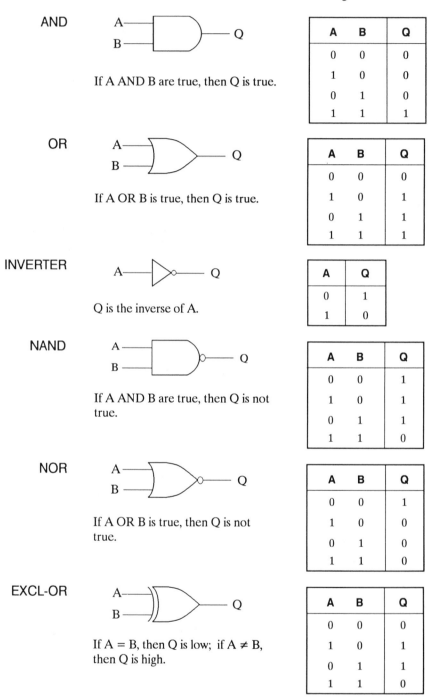

AND

If A AND B are true, then Q is true.

A	B	Q
0	0	0
1	0	0
0	1	0
1	1	1

OR

If A OR B is true, then Q is true.

A	B	Q
0	0	0
1	0	1
0	1	1
1	1	1

INVERTER

Q is the inverse of A.

A	Q
0	1
1	0

NAND

If A AND B are true, then Q is not true.

A	B	Q
0	0	1
1	0	1
0	1	1
1	1	0

NOR

If A OR B is true, then Q is not true.

A	B	Q
0	0	1
1	0	0
0	1	0
1	1	0

EXCL-OR

If A = B, then Q is low; if A ≠ B, then Q is high.

A	B	Q
0	0	0
1	0	1
0	1	1
1	1	0

Figure 2.2. Circuit symbols and truth tables for six basic logic gates.

Figure 2.3. An AND application where one input is a control which determines whether
or not the signal at the other
input passes to Q.

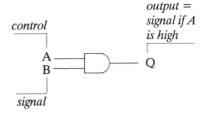

2.1.1 AND

An application of AND logic is to think of one of the inputs as a control and
the other as a "signal," such as a digital clock. If the control input A is high,
the output Q equals the signal input B, be it high or low or alternating between high and low. This can be seen by looking at input B and output Q on
the rows of the truth table where A is high. On the other hand, when control
is low, the output is low no matter what the signal is. This can be seen by
looking at the rows of the truth table where A is low. So the signal passes the
AND gate only when control is high. The idea is shown in Figure 2.3.

Consider the four-input AND gate in Figure 2.4. The output is high
only if all four inputs are high.

2.1.2 OR

A way to think about OR logic is that the output is low only if all inputs are
low. This helps visualize the operation of the circuit in Figure 2.5. With the
two inverters, the four inputs to the OR gate are all low only when the 4 bits
A, B, C and D are, in order, 0110. If any bit is different, at least one OR input
is high and therefore the output is high.

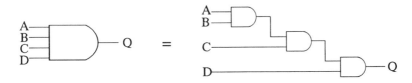

Figure 2.4. A 4-input AND gate implemented with three 2-input gates.

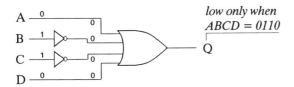

Figure 2.5. An OR based circuit.

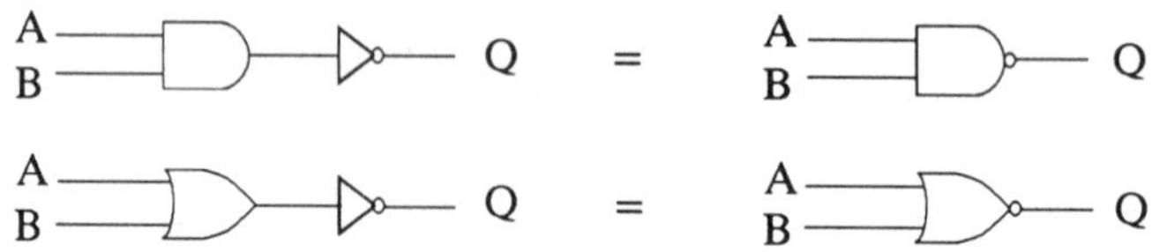

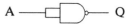

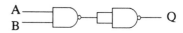

Figure 2.6. NAND and NOR gates are AND and OR gates inverted.

Figure 2.7. An INVERTER made from a NAND.

Figure 2.8. An AND made from two NAND's.

2.1.3 INVERTERS, NAND'S and NOR'S

NAND and NOR gates are identical to AND's and OR's with outputs inverted as shown in Figure 2.6. Similarly, the outputs of NAND's and NOR's sent through inverters produce AND's and OR's.

An inverter can be created from a NAND gate by connecting the two inputs together as shown in Figure 2.7. The operation is verified by just looking at the respective rows of the truth table where both A and B are the same. The circuit in Figure 2.8 shows an AND gate made from two NAND's.

Consider the circuit in Figure 2.9. The expanded truth table shows that the circuit carries out OR logic. An INVERTER added to the output would produce NOR logic. So, in addition to INVERTER and AND gates, OR's and NOR's can be made from NAND's. Since the flip-flop, counter and decoder/demultiplexer circuits presented in later sections are all implemented with basic gates, "everything" can be built from NAND's. Hence, it's the most basic gate.

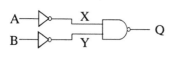

A	B	X	Y	Q
0	0	1	1	0
1	0	0	1	1
0	1	1	0	1
1	1	0	0	1

Figure 2.9. Circuit and truth table for an OR gate made from a NAND and INVERTERS.

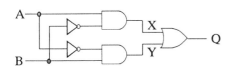

A	B	X	Y	Q
0	0	0	0	0
1	0	1	0	1
0	1	0	1	1
1	1	0	0	0

Figure 2.10. An EXCLUSIVE-OR circuit and explanatory truth table.

2.1.4 EXCLUSIVE-OR

EXCLUSIVE-OR logic is implemented by the circuit in Figure 2.10. The truth table shows the intermediate states X and Y for all possible inputs. The final output is deduced by applying X and Y to the OR gate. The output is low when A and B are equal and high when they are not equal.

2.2 Integrated Circuits, Breadboarding and Troubleshooting

Digital circuits are thought about in terms of combinations of the symbols in Figure 2.2, but they are built using integrated circuits. The following five IC's contain the basic gates needed for this book.

7404 contains six independent INVERTER gates
7408 contains four 2-input AND gates
7432 contains four 2-input OR gates

7421 contains two 4-input AND gates
7425 contains two 4-input NOR gates (1G and 2G must be high)

Three additional IC's are introduced.

7400 contains four 2-input NAND gates
7402 contains four 2-input NOR gates
7486 contains four 2-input EXCL-OR gates

Pin assignments for all eight chips are given in Figure 2.11 and Appendix C.

Integrated circuits are implemented in a variety of logic "families" and "subfamilies," each with its own electrical characteristics. In this book, only TTL (transistor transistor logic) chips are used. The differences among TTL subfamilies (e.g., 7404, 74LS04, 74H04), while significant in engineering commercial products, are not important for the circuits in this book. However, popular LS chips are recommended, if available. Appendix A gives references which describe logic families and their characteristics.

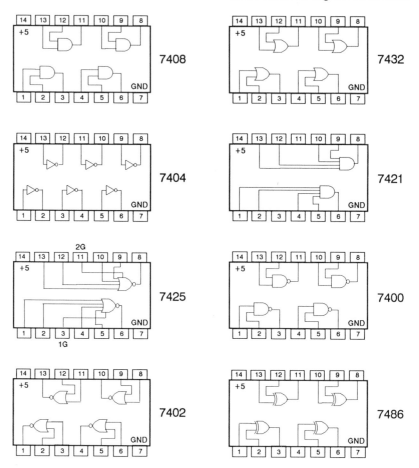

Figure 2.11. Pin assignments for eight basic gate integrated circuits.

2.2.1 Breadboard Systems

In order to build and test circuits in this and later chapters, a bread-board system is required. The E&L Instruments DD-1, shown in Figure 2.12, contains a +5 volt power supply, logic switches, pulsers, digital clock, light emitting diodes (LED's) and a breadboard on which IC's are wired. Chips are traditionally oriented so pin 1 is on the bottom left as shown.

A system LED is "lit" when connected to a high output. It is not lit when the output is low or disconnected. A logic probe distinguishes between disconnected and low and is an essential tool for troubleshooting.

Appendix B lists vendors for breadboard systems and logic probes.

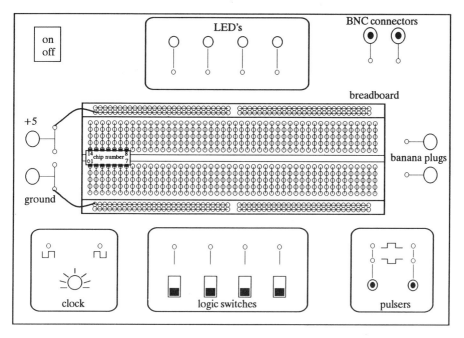

Figure 2.12. Layout of a typical breadboard system. IC's are inserted with the chip label as shown, the "notch" on the left and/or a circle above pin 1.

2.2.2 Circuit Construction

As an illustration of circuit construction and testing, recall the diagram for EXCLUSIVE-OR logic in Figure 2.10. Figure 2.13 shows the circuit implemented with 7404, 7408 and 7432 chips. When constructed on a breadboard, don't forget to connect +5 volts and ground to all IC's. Inputs A and B may originate with breadboard system logic switches and the output goes to an LED or logic probe. Circuit operation is verified by observing the output for all possible combinations of inputs.

2.2.3 Troubleshooting

Perhaps the most important skill in constructing and testing digital circuits is figuring out what's wrong when the anticipated operation is not observed. The key to troubleshooting is thoroughly understanding the circuit.

Possible problems fall into three categories: the breadboard system itself, wiring errors and defective chips and faulty logic design.

The breadboard system is checked by first verifying all +5 and ground connections. The logic probe should be put directly on IC pins thus identify-

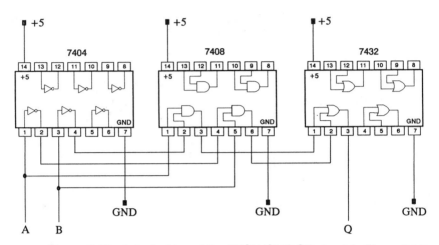

Figure 2.13. Actual wiring of the EXCLUSIVE-OR circuit in Figure 2.10.

ing possible bent pins, bad connections and even a defective breadboard. Also, all inputs originating with system switches and pulsers should be checked.

Finding wiring errors begins with being sure chips are not upside down. Next, check the outputs of all gates with the logic probe. This requires a clear picture of what each gate should do. The cause of an incorrect output may be the chip originating the output or whatever it's connected to. Outputs with incorrect values should be disconnected to isolate the cause. Defective chips are rare but cannot be discounted.

After the breadboard system and wiring have been verified, the circuit's logic must be reexamined. Usually, in the course of checking the wiring, design flaws are realized.

The approach here applies specifically to basic gate circuits such as the example in Section 2.2.2. Additional troubleshooting pointers are given throughout the book.

2.2.4 Two Rules for Circuit Construction

When IC gates are wired to make a circuit, the following rules must be followed.

Two outputs must never be connected together. After all, if one output is high and the other is low, what is the connected value? Exceptions are the tri-state outputs discussed in Section 2.10.

An output may, however, be connected to multiple inputs. There are electrical restrictions on how many inputs a single output can support. This is called "fanout" and varies with logic family. A good rule of thumb is to "buffer"

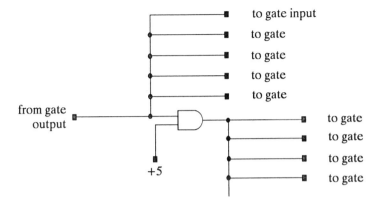

Figure 2.14. Outputs connected to more than six inputs should be "buffered." One approach is with an AND gate.

outputs which are connected to more than six inputs. Figure 2.14 shows how to do this with an AND gate.

2.3 Sequential Logic and Digital Clocks

The basic gates presented in Section 2.1 are sufficient to construct a variety of worthwhile circuits, including arithmetic and decoding. Truth tables are used to describe circuit operations. A common characteristic is that when any one or several inputs change, the output becomes the one specified by the "new" row of the table. The previous sequence of input changes never matters. While this may seem trivial, it's not. There exists an entirely different category of logic operations where the effect of an input change is very much determined by earlier changes. For example, when the input to a counter changes, the count value increments. But the new value depends on the previous value which itself depends on the number of earlier input changes. The flip-flop presented in the next section is the main building block of circuits where sequence matters.

Sequencing events in time is the backbone of measurement systems and accurate digital clocks are the backbone of sequences. Crystal oscillators with TTL outputs are used for all timing operations in this book. The oscillators come in integrated circuit packages and run off +5 volts and ground. Available frequencies are 1.0000, 2.0000 and 4.0000 MHz and higher. Both logic and pin assignment drawings are given in Figure 2.15. Vendors are listed in Appendix B.

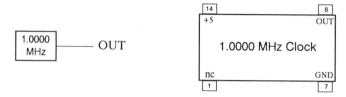

Figure 2.15. Logic drawing and pin assignments for a TTL compatible digital clock oscillator.

Signal generators with TTL outputs and breadboard system clocks based on 555 timers are useful in testing circuits. However, the stability, accuracy and ease of use of crystal clocks make them the preferred choice for all applications which require hardware timing.

In sequential logic, timing diagrams replace truth tables as the major way to think about and explain operations. Figure 2.16 is the diagram for a 1.0000 MHz clock. Note that the signal is high for half a cycle and low the other half. ΔT is the period or time for a cycle and equals $1.0000\,\mu$sec.

A simple clock application, carried out by the circuit in Figure 2.17, counts the number of 1.0000 MHz cycles while a control pulse is high. The timing diagram is in Figure 2.18. If the count value is initially zero, after the sequence, the number of cycles counted multiplied by the clock period (1.0000 μsec) equals the time the pulse was high.

A problem with the example is that the anticipated times to be determined must match the clock frequency and counter capacity. For instance, suppose the control pulse is high 1.0 ms (.0010 sec). Counter capacity must then be more than 1000, since during 1.0 ms there are 1000 cycles of a 1 MHz clock. A reasonable system requires a variety of lower clock frequencies.

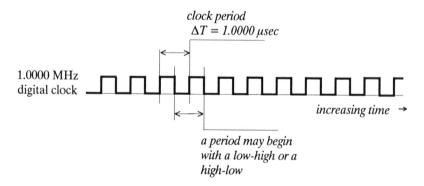

Figure 2.16. Timing diagram for a 1.0000 MHz digital clock.

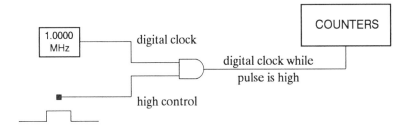

Figure 2.17. A time measurement circuit in which the counters receive the clock signal only when the control pulse is high.

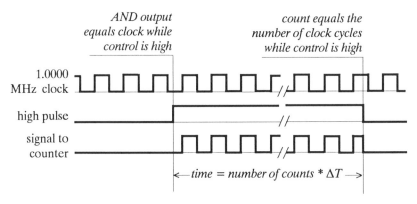

Figure 2.18. Timing diagram for counter/timer circuit in Figure 2.17.

The major focus of the remainder of this chapter and Chapter 6 is powerful and flexible ways to produce lower but still accurate and stable clock frequencies.

Sequential circuits are difficult to troubleshoot. While most logic probes indicate when a digital signal is pulsing, they cannot verify either the logic sequence or frequency. While digital frequency meters easily determine frequency, they cannot verify a logic sequence. For instance, in the example above, a meter can check the 1.0000 MHz clock frequency but cannot analyze the signal going to the counter. An oscilloscope is required to test logic sequences. But, to produce stable patterns, scopes require continuously repeating signals. A method to test the sequence in the example is shown in Figure 2.19. A breadboard system clock is set between 10 and 100 KHz and replaces the control pulse at the AND gate. Counts should pass the AND gate during the breadboard clock's high half cycle and repeat every cycle. If the breadboard clock goes to one scope channel and is used for triggering, the count signal is easily displayed on a second channel and the logic can be

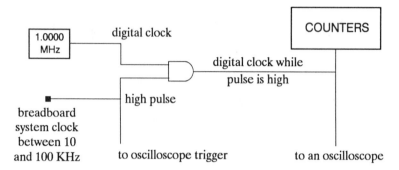

Figure 2.19. Troubleshooting the counter/timer circuit with an oscilloscope.

verified. While this case is straightforward, the approach is applicable to circuits with complex sequences.

Memory scopes avoid the need for repetition and therefore the breadboard system clock, but only expensive models work at 1 MHz frequencies.

2.4 Flip-flops

2.4.1 SET/CLEAR Flip-flop

The flip-flop is a major building block for sequential circuits. The version shown in Figure 2.20 has two control inputs and one output Q. The digital state of Q is the current "value" of the flip-flop: if high, the flip-flop is SET to 1; if low, the flip-flop is CLEARED to 0. The inputs, SET* and CLEAR*, are ordinarily inactive. SET*'s activation sets the flip-flop and CLEAR*'s activation clears the flip-flop. Obviously, SET* and CLEAR* must never act simultaneously. At any time, the value of the flip-flop is a record of whether SET* or CLEAR* was most recently activated.

Figure 2.20. Logic drawing for a set/clear flip-flop.

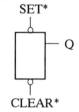

Timing diagrams are the standard tool in thinking about and explaining flip-flops. The set/clear operation is shown in Figure 2.21. Because SET* and CLEAR* act when briefly taken low, they are said to be "active-low."

Throughout this book, the names of active-low control signals have appended asterisks. Further, in logic diagrams active-low inputs are designated by a "bubble." In thinking about circuits, it's important to automatically picture a control signal whose name has an appended "*" as ordinarily high and acting when taken low. The same thing applies to "bubble" inputs in logic diagrams.

Active-low control signals are sometimes designated by a bar over the name or by the character "-" preceding the name.

The operation of the flip-flop described by the timing diagram in Figure 2.21 is:

(1) When and while SET* is active, Q = 1. Nothing happens when SET* returns inactive.
(2) When and while CLEAR* is active, Q = 0. Nothing happens when CLEAR* returns inactive.
(3) SET* and CLEAR* must never both be active.

2.4.2 Flip-flop Implementation with NAND Gates

How is flip-flop logic achieved? While not important for the purposes of this book, the implementation is a worthwhile exercise in basic gate circuit analysis and it further demonstrates the idea that complex digital circuits are nothing more than combinations of basic gates. The circuit in Figure 2.22 carries out the set/clear operation. In order to understand the logic, truth tables for both NAND gates, labeled with signal names, are given. Table rows are designated a, b, c and d.

The circuit is analyzed by starting out with SET* and CLEAR* high. From the NAND truth tables, Q1 and Q2 still cannot be determined because

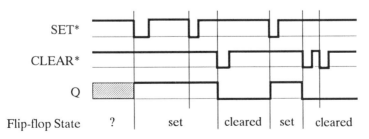

Figure 2.21. Timing diagram which illustrates the set/clear operation. The initial flip-flop value Q is undetermined (?).

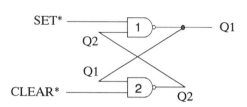

NAND-1

row	SET*	Q2	Q1
a	0	0	1
b	1	0	1
c	0	1	1
d	1	1	0

NAND-2

row	Q1	CLR*	Q2
a	0	0	1
b	1	0	1
c	0	1	1
d	1	1	0

Figure 2.22. Implementation of a set/reset flip-flop with NAND gates.

a single high input does not fix NAND outputs. However, the following sequence does.

(1) Take SET* low. This puts NAND-1 on truth table row a or c and makes Q1 definitely high (the flip-flop is set). With Q1 and CLEAR* high, NAND-2 is on row d and Q2 is low. This puts NAND-1 on row a and Q1 remains high.

(2) Take SET* back high. This moves NAND-1 from row a to row b and keeps Q1 high. NAND-2 stays on row d.

(3) Take CLEAR* low. This moves NAND-2 from row d to b and Q2 from low to high. A high Q2 takes NAND-1 from row b to d and Q1 goes low (the flip-flop Q is cleared). A low Q1 takes NAND-2 from row b to a and Q2 stays high.

(4) Return CLEAR* high. This takes NAND-2 from row a to c causing no change in Q2. NAND-1 stays on row d and Q1 is unchanged.

As long as SET* and CLEAR* never act simultaneously, SET* makes Q1 high and CLEAR* makes Q1 low.

2.4.3 7474 Dual, Positive-edge Triggered, D-type Flip-flop

The 7474 is the only flip-flop used in this book. In addition to PRESET* (the same as SET*) and CLEAR*, there are CLOCK and DATA inputs. And, in addition to Q, Q* (the complement of Q) is produced. The 7474 logic symbol and pin assignments are in Figure 2.23. The chip has two independent flip-flops (hence the label dual).

The PRESET* and CLEAR* controls work identically to SET* and CLEAR* above.

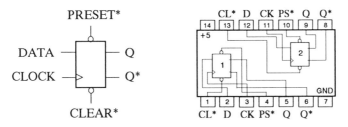

Figure 2.23. Logic symbol and IC pin assignments for the 7474 flip-flop.

The new feature is a clock/data mode. If PRESET* and CLEAR* are inactive, when the signal at the CLOCK input goes from low to high (a positive edge), Q is made equal to the signal at the DATA input. Put another way, CLOCK going high loads DATA into the flip-flop. If either PRESET* or CLEAR* is active, CLOCK going high has no effect.

The timing diagram in Figure 2.24 illustrates both preset/clear and clock/data operations. PRESET* and CLEAR* act while low, whereas CLOCK acts only when a low-high transition occurs. The applications in the next three sections illustrate the two modes as well as logic design with flip-flops.

2.4.4 4-bit Memory

Consider the circuit in Figure 2.25. The PRESET* and CLEAR* controls of all four flip-flops are wired inactive. Therefore, the only time flip-flop values can change is when the signal LOAD (connected to the four CLOCK's) goes high. At those times, FF-1 is made equal to the signal D1, FF-2 to D2, FF-3

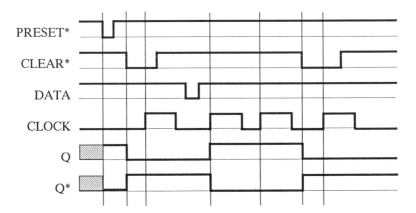

Figure 2.24. Timing diagram illustrating 7474 operation. The vertical lines indicate when actions occur.

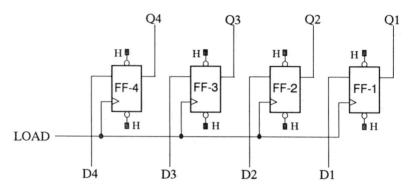

Figure 2.25. A 4-bit memory register built from 7474 flip-flops.

to D3 and FF-4 to D4. The flip-flops hold the D values until the next LOAD low-high. Hence, the circuit acts as a 4-bit memory register.

2.4.5 Toggling

In addition to being the basis for memory, flip-flops are the building block of another digital operation: clock frequency reduction. Consider the circuit in Figure 2.26. PRESET* and CLEAR* are wired inactive. The IN signal goes to the CLOCK input; OUT is the flip-flop's value Q; Q*, the complement of OUT, is the DATA input. Therefore, when IN goes high, the flip-flop is made equal to Q*, the complement of its current value. OUT reverses state. The process is called toggling and works the same as a toggle switch: if an activation turns lights off, the next activation turns them on, the next off again, etc.

The timing diagram in Figure 2.27 analyzes the operation. The time scale is greatly expanded so transitions do not appear instantaneous. The gate delay between when the flip-flop's CLOCK input goes high and the outputs change is also shown. Manufacturers specify this time for a 74LS74 chip as typically 13 ns. For the case shown in the diagram, when IN goes high, the flip-flop is high and DATA is low. After the gate delay, the flip-flop goes low and Q* high. OUT reverses and the stage is set for the next IN low-high to again reverse OUT.

Figure 2.26. A toggling 7474 flip-flop where OUT reverses every time IN goes high.

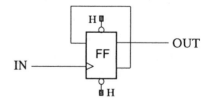

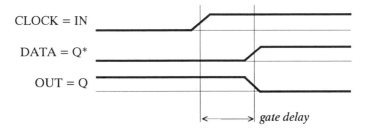

Figure 2.27. Timing diagram for the toggling operation showing the 7474 gate delay. The time scale is greatly expanded.

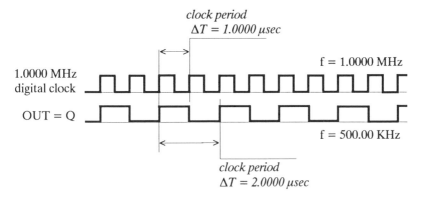

Figure 2.28. Timing diagram showing how a toggling flip-flop divides a clock frequency by 2.

The timing diagram in Figure 2.28 shows the toggle operation when the CLOCK signal is the 1.0000 MHz digital oscillator. The output is also a clock with twice the input clock's period. This makes the frequency 1/2 the input's or 500.00 KHz. Additional flip-flops could further divide the frequency as desired.

Section 2.7 shows how multiple toggling is the basis for digital counters.

2.4.6 Synchronization

A common problem in measurement and control circuits is synchronization between a START signal and a continuously running CLOCK. A typical design problem is shown in Figure 2.29. The Timing Signal (TS) is initially low but becomes equal to CLOCK after START goes high. Since START may go high at any point in a CLOCK cycle, it's usually necessary to delay TS until the beginning of the next full CLOCK cycle. The circuit in Figure 2.30 accomplishes this.

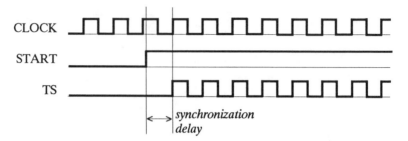

Figure 2.29. Timing diagram showing synchronization between START and CLOCK signals so that TS is delayed until the beginning of the next clock cycle.

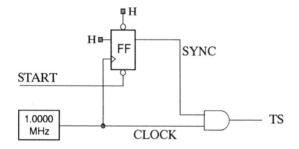

Figure 2.30. A circuit which implements the synchronization in Figure 2.29's timing diagram.

While START is low, the flip-flop (SYNC) is low and CLOCK does not pass the AND gate. When START goes high, the flip-flop stays low. However, the next CLOCK low-high (the beginning of the next full CLOCK cycle) sets the flip-flop. When SYNC goes high, TS equals CLOCK as desired.

2.5 Electrical Properties of TTL Integrated Circuits

Occasionally, it's necessary to deal with the electrical properties of TTL integrated circuits. This section covers: 1) the relationship between outputs and inputs; 2) the reason for the predominance of active-low controls; 3) wiring light-emitting diodes (LED's) and digital switches; 4) recognition of "noise" and "glitch" problems; and 5) gate delays. References in Appendix A give complete treatments of the implementation of TTL gates with transistors and diodes and therefore the basis for the properties presented below.

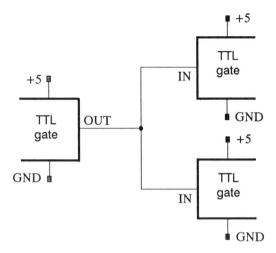

Figure 2.31. The output of a TTL gate connected to two inputs. When OUT is high no current flows; when OUT is low, current flows from the inputs to the output.

2.5.1 TTL Inputs and Outputs

Figure 2.31 shows the output of a TTL gate connected to the inputs of two gates. When the output is +5 volts, little current flows either way between the output and inputs. However, when the output is 0 volts, current flows from a higher electric potential at the inputs to the zero potential at the output. Therefore, a low output must receive or sink current.

The number of inputs to which an output can be connected is determined by the total current which can be sinked. Fanout, as stated earlier, varies with logic family.

A disconnected input behaves as if a high voltage were applied. Therefore, logic gates work as if any disconnected inputs are high.

Most control signals in digital circuits are active-low. This means disconnected controls are inactive. For instance, computer expansion slots have individual active-low controls which indicate when a board is installed. The controls in empty slots are disconnected and therefore are seen as inactive.

2.5.2 Wiring LED's to TTL Gates

Although not explicitly used in this book, LED's are worthwhile in indicating the status of logic signals (if they are not changing too fast). A diode is shown in schematic form in Figure 2.32a. When current is made to flow through the diode in the direction suggested by the "arrow," some of the electrical energy is converted to light. When no current is flowing, the diode is not lit. The question is how to wire LED's to TTL outputs.

Consider a diode connected to the output of a TTL gate as shown in Figure 2.32b. When the gate is high, a current should flow and the diode light as desired. However, in the TTL scheme a high output cannot supply current. Therefore, the more complex circuit shown in Figure 2.32c is required. A high gate output produces a low inverter output and current flows from +5 through the diode to the inverter and on to ground. A low gate output produces a high inverter and no current flows between high and +5. The purpose of the resistor is to limit the current when the inverter output is low. Typical R values are 200 ohms. Although not required, the 7406 is a special inverter chip especially suited to running LED's.

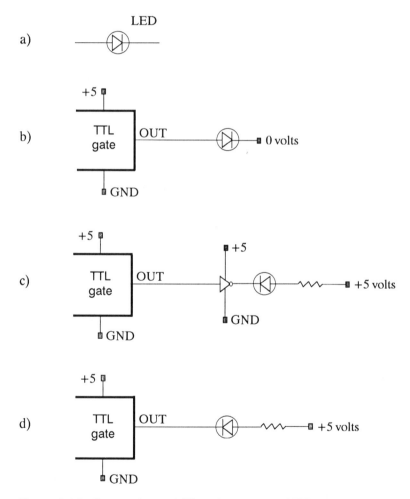

Figure 2.32. Connecting an LED to the outputs of TTL gates.

Figure 2.33. A digital switch, wired as shown, produces a high output when open and a low output when closed.

Sometimes it doesn't matter whether an LED displays the actual output or its complement. In those cases, the diode may be connected as shown in Figure 2.32d. When OUT is low, current flows into the gate and the LED lights.

2.5.3 Wiring Logic Switches

Figure 2.33 shows a logic switch connected to +5 and ground. An open switch produces a high output. (As stated earlier, when a high out is connected to one or several inputs, no current flows. Therefore, there is no voltage drop across the resistor and OUT = high.) When the switch is closed, the output is grounded and current flows from +5 through the resistor and closed switch to ground. The purpose of the resistor is to avoid shorting the +5 power supply when the switch is closed. R values range from 1 to 5 KΩ.

Figure 2.34 is a dip-switch/decoder circuit which carries out a useful operation. When the logic states of the four inputs, Y_A, Y_B, Y_C and Y_D, are the same as the corresponding outputs of the four independent switches, OUT is low. When any difference exists between an input and switch, at least one EXCLUSIVE-OR output is high which makes OUT high. For the case shown, S1 is low, S2 high, S3 low and S4 high. Therefore, the outputs of the EXCLUSIVE-OR's are all low (and the OR output is low) only if $Y_A = 0$, $Y_B = 1$, $Y_C = 0$ and $Y_D = 1$.

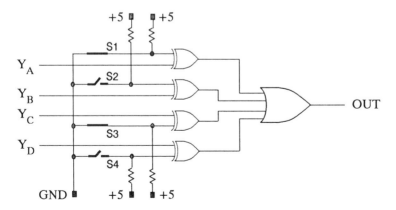

Figure 2.34. A four switch decoder circuit which produces a low OUT only when the four EXCLUSIVE-OR gates have low outputs.

2.5.4 Noise

Unwanted noise is present in all digital circuits, especially those with high
frequency clocks. Of the chips presented so far, flip-flops are particularly
sensitive. For example, suppose a flip-flop is set. All it takes is a brief low
noise spike on the CLEAR* control to inadvertently clear the flip-flop.

It's important to recognize when noise is causing problems. An approach is to put a circuit in some sort of repetitive loop, perhaps using the
breadboard system's clock. Then an oscilloscope is used to simultaneously
observe the output and the input suspected of causing trouble. If a problem
is found, a possible solution for active-low controls (such as SET* and CLEAR*)
is to put a .01μf capacitor between the control and ground.

It turns out that noise problems are more likely when digital circuits are
interfaced to computers. This is particularly so when control signals are
computer originated. In the next chapter, after parallel ports and associated
software are introduced, troubleshooting noise is presented in Section 3.5.6.

2.5.5 Glitches

Glitches are another problem in digital circuits. Unlike noise, glitches are
briefly occurring unintended states which, for instance, cause a false decode.
Suppose the four input signals (Y_A, Y_B, Y_C and Y_D) in the circuit in Figure
2.34 change from 1111 to 0000. Since the circuit decodes $Y_1 = 0$, $Y_2 = 1$, Y_3
$= 0$ and $Y_4 = 1$, OUT should remain high during and after the transition.
But what if the four-bits do not change simultaneously on a nanosecond time
scale. In that case, the following scenario is possible:

time	Y_A	Y_B	Y_C	Y_D
start	1	1	1	1
after 20 ns	1	1	0	1
after 30 ns	0	1	0	1
after 40 ns	0	0	0	1
after 50 ns	0	0	0	0

The EXCLUSIVE-OR gates may see 0101 long enough to produce a brief
low output.

Obviously, glitches must be avoided. As parallel ports and measurement circuits are presented in later chapters, the potential for glitches is carefully analyzed.

2.5.6 Gate Delays

Gate delays were touched on earlier. When one or more inputs change,
there is always a delay before the outputs change. For example, according to

manufacturers of 7404 inverters, the time between a high-to-low input change and the resulting output change varies as follows:

chip	delay times (ns) typical	maximum
7404	12	22
74H04	6	10
74L04	35	60
74LS04	9	15

Also, the time between a low-to-high input change and the resulting output transition varies as follows:

chip	delay times (ns) typical	maximum
7404	8	15
74H04	6.5	10
74L04	31	60
74LS04	10	15

2.6 Hexadecimal, Binary and Decimal Numbers

Digital signals are either 0 or 1. Often, several inputs and/or outputs combine to form 4-, 8- and 16-bit binary values. Sometimes the values represent numbers but they may also be instruction codes and command words. No matter which case, it's necessary to think about multi-bit digital signals in terms of binary, decimal and hexadecimal values. The purpose of this section is to show how to convert among the three number systems.

2.6.1 Decimal, Hexadecimal and Binary

Decimal is a base 10 number system, binary is base 2 and hexadecimal is base 16. The 10 decimal, 2 binary and 16 hex digits are:

decimal 0 1 2 3 4 5 6 7 8 9

binary 0 1

hex 0 1 2 3 4 5 6 7 8 9 A B C D E F

In all systems, numbers consist of digits in places. Positive integers (of interest here) start with the $base^0$ place and increase to the left ($base^1$, $base^2$, etc.). The places are:

decimal	$10^3(1000)$	$10^2(100)$	$10^1(10)$	$10^0(1)$
binary	$2^3(8)$	$2^2(4)$	$2^1(2)$	$2^0(1)$

hex $16^3(4096)$ $16^2(256)$ $16^1(16)$ $16^0(1)$

In counting, when the highest digit in a given place is reached, one is carried to the next higher place. The table below illustrates counting in all three systems.

binary	hex	decimal	binary	hex	decimal
0000 0000 0000	0 0 0	0 0 0 0	0000 1111 1010	0 F A	0 2 5 0
0000 0000 0001	0 0 1	0 0 0 1	0000 1111 1011	0 F B	0 2 5 1
0000 0000 0010	0 0 2	0 0 0 2	0000 1111 1100	0 F C	0 2 5 2
0000 0000 0011	0 0 3	0 0 0 3	0000 1111 1101	0 F D	0 2 5 3
0000 0000 0100	0 0 4	0 0 0 4	0000 1111 1110	0 F E	0 2 5 4
0000 0000 0101	0 0 5	0 0 0 5	0000 1111 1111	0 F F	0 2 5 5
0000 0000 0110	0 0 6	0 0 0 6	0001 0000 0000	1 0 0	0 2 5 6
0000 0000 0111	0 0 7	0 0 0 7	0001 0000 0001	1 0 1	0 2 5 7
0000 0000 1000	0 0 8	0 0 0 8	0001 0000 0010	1 0 2	0 2 5 8
0000 0000 1001	0 0 9	0 0 0 9	0001 0000 0011	1 0 3	0 2 5 9
0000 0000 1010	0 0 A	0 0 1 0	1111 1111 1100	F F C	4 0 9 2
0000 0000 1011	0 0 B	0 0 1 1	1111 1111 1101	F F D	4 0 9 3
0000 0000 1100	0 0 C	0 0 1 2	1111 1111 1110	F F E	4 0 9 4
0000 0000 1101	0 0 D	0 0 1 3	1111 1111 1111	F F F	4 0 9 5
0000 0000 1110	0 0 E	0 0 1 4			
0000 0000 1111	0 0 F	0 0 1 5			
0000 0001 0000	0 1 0	0 0 1 6			
0000 0001 0001	0 1 1	0 0 1 7			
0000 0001 0010	0 1 2	0 0 1 8			
0000 0001 0011	0 1 3	0 0 1 9			
0000 0001 1101	0 1 D	0 0 2 9			
0000 0001 1110	0 1 E	0 0 3 0			
0000 0001 1111	0 1 F	0 0 3 1			
0000 0010 0000	0 2 0	0 0 3 2			
0000 0010 0001	0 2 1	0 0 3 3			
0000 0010 0010	0 2 2	0 0 3 4			

2.6.2 Conversion of Binary and Hexadecimal to Decimal

A decimal number $d_3d_2d_1d_0$ can be written out as follows:

decimal = d_3*10^3 + d_2*10^2 + d_1*10^1 + d_0*10^0

where d_i is the digit in the 10^i place. For example

2904 = 2*1000 + 9*100 + 0*10 + 4*1

The same process works in converting binary and hex to decimal. Consider the binary number $b_4b_3b_2b_1b_0$ where b_i is the digit (0 or 1) in the 2^i place. The decimal value is:

decimal = b_4*2^4 + b_3*2^3 + b_2*2^2 + b_1*2^1 + b_0*2^0

For example, 11001 converts as:

decimal = 1*16 + 1*8 + 0*4 + 0*2 + 1*1 = 25

Consider the hex number $h_2h_1h_0$ where h_i is the digit (0 to F) in the 16^i place. The decimal value is:

```
decimal = h₂*16² + h₁*16¹ + h₀*16⁰
```

$$decimal = h_2*16^2 + h_1*16^1 + h_0*16^0$$

In doing the calculation, the decimal values of the hex digits A to F are used as well as the decimal values of the 16^i's. For example, C07 converts as:

```
value = 12*256 + 0*16 + 7*1 = 3079
```

(the decimal value of C is 12).

2.6.3 Conversion of Decimal to Hexadecimal

Occasionally, it's necessary to determine the hex value of a decimal number. The conversion is illustrated below for 28,945. The goal is to find the hex digits such that:

$$28945 = h_3*16^3 + h_2*16^2 + h_1*16^1 + h_0*16^0$$

or $\quad 28945 = h_3*4096 + h_2*256 + h_1*16 + h_0*1$

The value starts in the h_3 place with:

$$h_3 = INT(28945/4096) = 7$$

where INT means the quotient rounded off to the nearest lower integer.

$$h_2 = INT[(28945 - 7*4096)/256] = 1$$

$$h_1 = INT[(28945 - 7*4096 - 1*256)/16] = 1$$

$$h_0 = 28945 - 7*4096 - 1*256 - 1*16 = 1$$

So 28,945 decimal equals 7111 hex.

2.6.4 Binary-hex and Hex-binary Conversion

By far the most frequent conversion is between hexadecimal and binary. It's efficient, especially in writing software, to work with the few hex digits required to specify a value. But when values become multi-bit digital signals, the format is necessarily binary.

The relationship between hex and binary is demonstrated in the counting table. Each 4 binary bits form a hex place, or each hex digit corresponds to 4 binary bits. Mathematically, this is because:

$$b_3*2^3 + b_2*2^2 + b_1*2^1 + b_0*2^0 = h_0*16^0$$

$$b_7*2^7 + b_6*2^6 + b_5*2^5 + b_4*2^4 = h_1*16^1$$

$$b_{11}*2^{11} + b_{10}*2^{10} + b_9*2^9 + b_8*2^8 = h_2*16^2$$

To convert from binary to hex, divide the binary number starting on the right into groups of 4 bits. The 16^0 hex digit is the value of the first 4 bits. The 16^1 digit is the value of the next four bits. And so on. For example:

        1011000111
    0010   1100   0111
      2   c   7
             2C7

To convert from hex to binary, the low 4 binary bits give the value of the h_0 hex digit. The next 4 binary bits give the value of the h_1 digit. And so on. For example:

          C09
    C   0   9
    1100   0000   1001
        110000001001

2.7 74192 and 74193 Counters

Counters are important in generating timing and control signals and they are used directly for data acquisition. Section 2.3 included a counter circuit which measures time. Another application is a scaler where, for instance, counters determine the number of radioactive decays detected in, say, 1.0000 second. In addition to actually taking data, counters produce from a reference clock the 1 second pulse during which measurement occurs.

The purpose of this section is to present the count process, to show how it's implemented with flip-flops, to describe the counter IC's used in this book, and to present two applications.

2.7.1 Digital Counting

Consider the counter shown in logic form in Figure 2.35. The four outputs DCBA specify the binary count value. Every time the input goes from low-

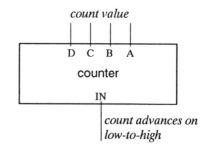

Figure 2.35. Logic drawing
of a 4-bit binary counter.

to-high, the count advances by one. The outputs undergo the following pattern.

decimal	D	C	B	A
0	0	0	0	0
1	0	0	0	1
2	0	0	1	0
3	0	0	1	1
4	0	1	0	0
5	0	1	0	1
6	0	1	1	0
7	0	1	1	1
8	1	0	0	0
9	1	0	0	1
10	1	0	1	0
11	1	0	1	1
12	1	1	0	0
13	1	1	0	1
14	1	1	1	0
15	1	1	1	1
0	0	0	0	0

As can be seen from the A column, every time the count advances, bit A reverses (toggles). Bit B toggles every time bit A goes low; bit C toggles every time bit B goes low; and bit D toggles every time bit C goes low. Figure 2.36 presents the count timing pattern. Although not required, COUNTS-IN is a periodic clock. Also, after reaching 1111, the count rolls over to 0000.

The count operation is carried out by the flip-flop circuit in Figure 2.37. Every time COUNTS-IN (to FF-A's CK input) goes high, Q_A toggles as required. Every time Q_A goes low, Q_A^* goes high causing FF-B to toggle. Every time Q_B goes low, Q_B^* goes high causing FF-C to toggle. Finally, every time Q_C goes low, Q_C^* goes high causing FF-4 (Q_D) to toggle. Once started, $Q_D Q_C Q_B Q_A$ follows the count timing pattern.

Before the 7474 flip-flop was introduced, the implementation of set/reset with NAND gates was presented. So far, the count operation with 7474's

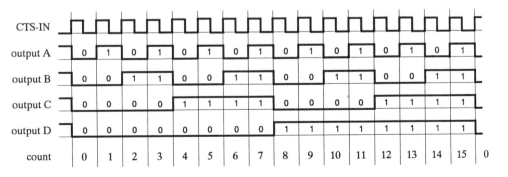

Figure 2.36. Timing diagram for a 4-bit binary counter.

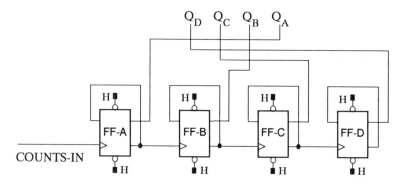

Figure 2.37. Flip-flop circuit which carries out the timing pattern in Figure 2.36.

(and therefore with NAND and other basic gates) has been described. A pair of counter IC's is next. Naturally, manufacturers design chips with as many features as possible.

2.7.2 74192 and 74193 Counters

There are two types of 4-bit counters: binary and decade. The former's outputs are a 4-bit number between 0000 and 1111. The latter's outputs are the binary value of a decimal digit between 0 and 9. Put another way, binary counters roll over from 1111 to 0000 and decade counters from 9 (1001) to 0 (0000). The 74192 is a general purpose, 4-bit decade counter and the 74193 is an identical binary counter. Figure 2.38 is a logic drawing of the chips. Eight inputs are on the bottom and six outputs on the top. $Q_D\ Q_C\ Q_B\ Q_A$ is the count value. The inputs are:

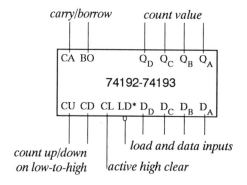

Figure 2.38. Logic drawing of the 73192 decade and 74193 binary counters.

CL	**CL**ear	Makes the count zero when and while high; must be low for the chip to count. (There's no "*" appended to CL, which means the clear action occurs on high.)
LD*	**L**oa**D***	When active, inhibits counting and makes the count equal $D_D\,D_C\,D_B\,D_A$.
$D_D...D_A$	**D**ata	Data inputs used by LoaD*.
CU	**C**ount-**U**p	When CLear is low and LoaD* is inactive, a low-high transition increments the count by 1.
CD	**C**ount-**D**own	When CLear is low and LoaD* is inactive, a low-high decrements the count by 1. If Count-Up is used, Count-Down must be left unconnected or high; and vice versa.

In addition to the count value, the outputs are:

CA	**CA**rry	Goes from low-to-high exactly when the count rolls over to zero. Therefore, when connected to a higher counter's Count-Up, causes it to increment. CArry goes low half-way between 9 and 0 in the 74192 and between 15 and 0 in the 74193. The CArry timing pattern is shown in Figure 2.39.
BO	**BO**rrow	Works the same as CArry except applies to counting down.

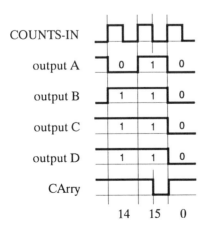

Figure 2.39. Timing diagram showing CArry for a 74193 binary counter. CArry goes high when a next higher counter should advance.

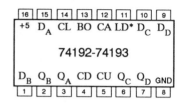

Figure 2.40. Pin assignment
diagram for 74192 and 74193
counters.

74192 and 74193 pin assignments are in Figure 2.40.

Figure 2.41 shows an 8-bit binary counter built from two 74193's. When
the low counter goes from 1111 to 0000, its CArry goes high. This causes the
next higher counter to increment as desired. Note the CLear inputs of both
counters are wired low. LoaD* and Count-Down are neither connected nor
used. If 74192's replaced the 74193's, the count range would be from 00 to
99.

2.7.3 Decoding Application

Consider an application with the following specification: count repeatedly
from 5 through 11 (0101 to 1011). Put another way, the instant the count
increments to 12 (1100), load 5 (1001). The sequence is:

5	0101
6	0110
7	0111
8	1000
9	1001
10	1010
11	1011
12 $\Rightarrow$ 5	1100 $\Rightarrow$ 0101

The sequence is implemented by generating a signal which is ordinarily

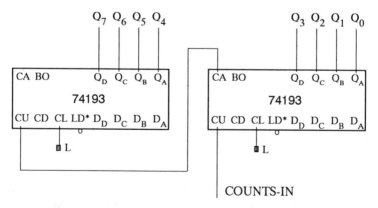

Figure 2.41. An 8-bit binary counter built from two 74193's.

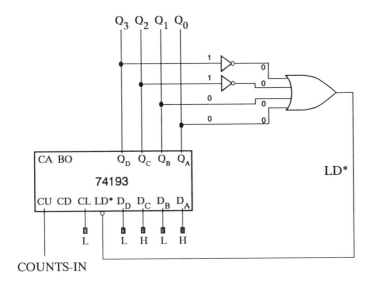

Figure 2.42. A circuit which counts repeatedly from 5 through 11.

high but briefly goes low when the count becomes 12 (1100). The circuit in Figure 2.42 accomplishes this. When the count goes from 11 (1011) to 12 (1100), for the first time in the sequence all four OR inputs are low and the signal LD* goes low. The instant LD* goes low, the count becomes $D_D D_C D_B D_A$ = 0101 (5). When this happens, the inputs to the OR gate are no longer all low and LD* returns high. Figure 2.43 shows the process on a greatly expanded time scale.

Four gate delays are shown in the timing diagram. Using manufacturers values for LS chips, the first delay is typically 30 ns. The second arises from the inverters and OR gate. With the circuit built with 74LS04 and 7425 chips, the net delay is around 28 ns. The third delay, again produced by the counter chip, is 25 ns. And the final delay by the decoder is again 28 ns. The entire sequence is typically 110 ns, and LD* is active around 50 ns.

The decoder part of the circuit has potential glitches. If all four counter outputs do not change simultaneously, LD* may be inadvertently activated. For instance, when the count goes from 7 (0111) to 8 (1000), a brief 1100 may occur. However, the 74192-74193 are "synchronous" counters where the outputs all change at the same time. Therefore, the circuit works with no potential for glitches.

The flip-flop counter in Section 2.7.1 is a "ripple" counter where the outputs do not simultaneously change.

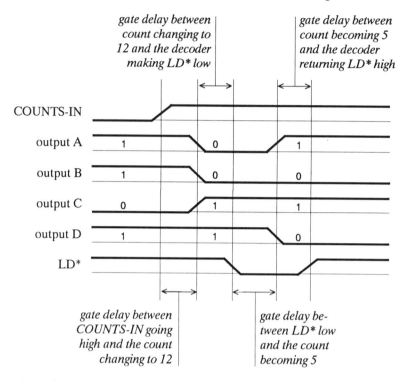

Figure 2.43. Timing diagram for the decoder part of the circuit in Figure 2.41. The time scale is greatly expanded to clarify gate delays.

2.7.4 100.00, 1000.0, 10.000K, 100.00K Clocks

The 1.0000 MHz digital oscillator described in Section 2.3 and four 74192 decade counters produce an accurate and stable set of digital clocks as shown in Figure 2.44. Since ten input changes produce a decade cycle, each Q_D output has exactly 1/10 the frequency of the CU input. Because Q_D is low when the count is between 0 and 7 and high only when the count is 8 and 9, these clocks signals are not symmetrically high and low.

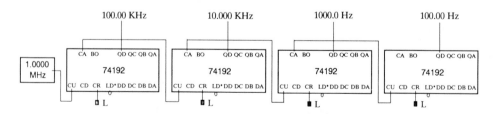

Figure 2.44. The generation of lower clock frequencies with counters.

2.8 74138 Decoder/demultiplexer

In interface and data acquisition circuits, it's often necessary to solve decoding problems such as the following. Consider these three controls (used in PIT circuits in Chapter 6): CS*, S/S and WR*. The goal is to generate a signal OUT* so that when and only when CS* is active and S/S is high, OUT* "follows" WR*. In all other cases, OUT* is inactive. Figure 2.45 shows the operation and Figure 2.46 gives its implementation. If, in the circuit, either CS* is inactive or S/S low, the intermediate signal CTRL* is high and OUT* is high no matter what WR* is. Conversely, OUT* "follows" WR* when CTRL* is low which requires a high S/S and a low CS*.

As long as decoding problems remain straightforward, circuits such as the one in Figure 2.46 are all that's needed. However, suppose it's necessary to generate three OUT*'s from CS_1*, CS_2* and CS_3* so that when CS_i* is active and S/S* is high, OUT_i* (only) follows WR*. Implementation is tedious with basic gates but not with the 74138 decoder/demultiplexer.

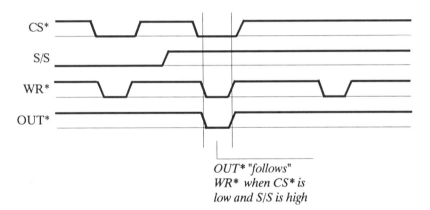

OUT "follows"*
WR when CS* is*
low and S/S is high

Figure 2.45. Specifications for a typical decoding problem.

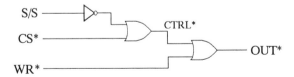

Figure 2.46. Implementation of the timing pattern in Figure 2.45. OUT* follows WR* when CTRL* is low. CTRL* is low when S/S is high and CS* low.

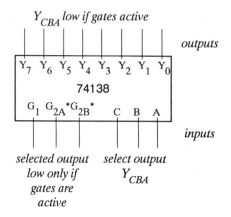

Figure 2.47. Annotated logic drawing of the 74138.

2.8.1 74138

An annotated 74138 logic drawing is given in Figure 2.47. The chip has six inputs and eight outputs (Y_0 to Y_7) which work as follows.

Inputs C B A	form a 3-bit binary number, with values ranging from 000 to 111 (0 to 7 decimal allowing eight unique values). The value of CBA selects the corresponding number output Y_{CBA}.
Outputs Y_0...Y_7	are almost always high. However, the selected output Y_{CBA} is low when the GATE inputs are "right." Since CBA can have only one value at a time, only one output can be selected at a time and therefore only one output can be low at a time. The objective of most 74138 applications is to make a selected output low but only when and while a set of several gate conditions is met.
Gate G_1	input must be active (high) for the chip to "work" (for the selected output to be low). Whenever G_1 is inactive, all outputs are high no matter what CBA is and no matter what the other gates are.
Gates G_{2A}^*, G_{2B}^*	must both be active (low) for the chip to "work." If either G_{2A}^* OR G_{2B}^* is inactive, all outputs are high, no matter what CBA is or what G_1 is.

In summary, the chip "works" only if all three gates are active.

74138 pin assignments are given in Figure 2.48 and reproduced in Appendix C.

Figure 2.48. 74138 pin
assignments.

2.8.2 A 74138 Application

How can a 74138 implement the second decoding case above? Specifically,
how can OUT_1^*, OUT_2^* and OUT_3^* be generated so OUT_i^* follows WR^*
when and while CS_i^* is active and S/S is high? The circuit in Figure 2.49
accomplishes the task. CS_1^*, CS_2^* and CS_3^* are connected to 74138 select
inputs A, B and C as shown. This makes the selected output correspond to
the CS^*'s as follows:

CS_3^*	CS_2^*	CS_1^*	selected
C	B	A	output
1	1	0	Y_6 (OUT_1^*)
1	0	1	Y_5 (OUT_2^*)
0	1	1	Y_3 (OUT_3^*)

The selected output goes low only when the gates are "right." For the case
shown, G_1 (S/S) must be high and both G_2^*'s (WR^*) must be low. Hence,
when S/S is high the selected OUT^* goes and remains low while WR^* is low.

2.8.3 A More Sophisticated 74138 Application

An important objective in digital circuits is to determine which of several
switches is active. Consider the case shown in Figure 2.50.

Suppose the four switches S_0^* to S_3^* are ordinarily high. This makes all
OR outputs high and therefore CTRL is high, no matter what the Y's are.

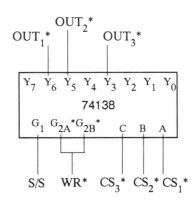

Figure 2.49. Implementation
of the decoding problem in
Section 2.8.2.

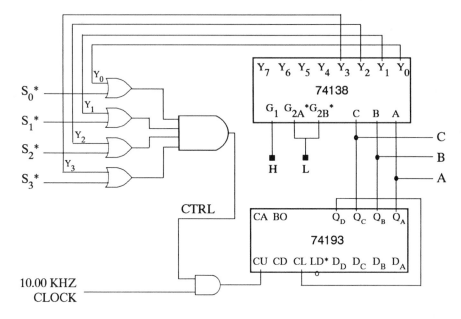

Figure 2.50. A switch encoding circuit. When a switch is active, outputs CBA equal the binary value of the switch's number.

When CTRL is high, the 10.00 KHz clock passes to the 74193 making it repeatedly count from 0 to 7. Note that bit Q_D is connected to the counter's active high CLear control. Bit Q_D is low as the count advances from 0 to 7. As soon as the count reaches 8, bit D goes high activating CLear which makes the count 0. The time bit D is high is determined by gate delays and is on the order of 25 ns. So the count repeatedly goes from 0 to 7 to 0 to 7, etc.

The low 3 bits of the count value are connected to the corresponding 74138 select inputs. Hence, the selected output always equals the count value between 000 (0) and 111 (7). All three gates are wired active so Y_{CBA} is low. This means output Y_0 is low while the count is 0, Y_1 while the count is 1, etc.

Each Y_i is ORed with the corresponding S_i^*. So if S_i^* is high when Y_i goes low (once each count cycle), the OR output is still high. The only way for any action to occur is for both S_i^* and Y_i to be low. Such a situation immediately makes CTRL low which in turn stops CLOCK and counting.

Suppose initially all switches are high. Then switch i is taken low. The next time the count becomes i, Y_i and CTRL go low and counting stops with the count value equal to i. Counting cannot resume until CTRL goes high which happens only when S_i^* returns high.

In essence, the circuit continually scans a set of switches and freezes on the first low switch. The outputs CBA hold the number of the active switch.

The circuit is easily expanded from four to eight switches. And with additional 74138's or more powerful specialized IC's, an encoding circuit which
produces an 8-bit code for up to 256 switches (or keys) is conceivable.

2.8.4 Troubleshooting Counters and 74138's

Suppose the switch encoding circuit of the last section is constructed and the
output (CBA) does not freeze on the number of an active switch. The following is a reasonable troubleshooting sequence.

First check CTRL with a logic probe. With a switch active, it should be
low. If not, disconnect it from the two-input AND gate and check again. If
still incorrectly high, check the OR gates and four-input AND first for wiring
errors and then for defective chips.

> *To make these checks, one must have access to IC pin assignments. While
> it's not necessary to have a detailed wiring diagram for circuits this simple,
> a good idea is to write actual chip pin numbers on the logic diagram. Don't
> forget to check +5 and ground on all chips.*

If CTRL is correct, use an oscilloscope to verify that CLOCK reaches
the counter (when no switch is active). If CLOCK does not pass the AND
gate, disconnect the AND output from CU and check again. If still absent,
check CLOCK's source and then the AND gate.

If CLOCK is reaching CU when no switch is active, the next step is to
verify the counter's operation. First, check that CLear is inactive low. Then
put CLOCK on oscilloscope channel 1 and successively Q_A, Q_B and Q_C on
channel 2. Q_A should be the toggle of CLOCK and so on. If any counter
output is incorrect, disconnect it from the 74138. If still incorrect, the counter
is either improperly wired or defective.

If the counter is working, the last thing to verify is the 74138 chip. First,
use a logic probe to check the gates. Then with CLOCK on scope channel 1,
successively look at the Y's on channel 2. Each should be low one CLOCK
period once every eight periods. If any Y is wrong, check the wiring and then
the 74138 chip.

Before all these steps are carried out, it's likely that the problem will be
identified.

2.9 74152 Data Selector/multiplexer

The 74152 multiplexer has a single output (W*), eight data inputs (D_7...D_0)
and three select controls (A, B and C). The binary value of CBA selects an
input and W* equals its inverse. Put another way, W* equals (D_{CBA})*. In

$$W^* = (D_{CBA})^*$$

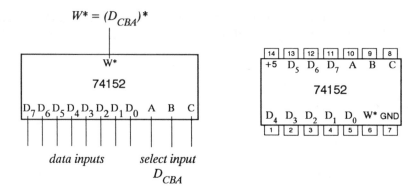

Figure 2.51. Annotated logic drawing and pin assignments for the 74152 multiplexer.

essence, the multiplexer connects one of many inputs to the single output. Figure 2.51 gives an annotated logic drawing and pin assignments.

The circuit in Figure 2.52 illustrates a multiplexer application. If START is initially low, FF-1 is set and CTRL is high. With CTRL high, AND-1 passes CLOCK-IN to the 74193's Count Up. However, because START's inverse is connected to the counter's CLear, the count is held at zero ($Q_D Q_C Q_B Q_A = 0000$). Also, because a low START clears FF-2, CLOCK-IN does not pass AND-2 and CLOCK-OUT is low.

Suppose an 8-bit value is supplied to the 74152's data inputs $D_7...D_0$. With the counter's $Q_C Q_B Q_A$ connected to CBA and equal to 000, W* equals D_0^* and SERIAL equals D_0. The interesting question is what happens when START goes high and the 74193 is free to count and both flip-flops respond to clock/data operations.

The next CLOCK-IN low-high after START goes high increments the count to 1 and SERIAL equals D_1. Then, each CLOCK-IN low-high advances the count and SERIAL is successively D_2, D_3, D_4, D_5, D_6 and D_7. When the count reaches 8, Q_D goes high which clears FF-1 and makes CTRL low. With CTRL low, AND-1 no longer passes CLOCK-IN and the 74193 stops counting (with the count 8 and $Q_C Q_B Q_A = 000$). At this point, SERIAL again equals D_0. Nothing else happens to SERIAL until START returns low and subsequently high.

In addition to SERIAL, the circuit produces CLOCK-OUT. Initially, when START is low, FF-2 is cleared, AND-2 stops CLOCK-IN and CLOCK-OUT is low. The first CLOCK-IN low-high after START goes high sets FF-2 and CLOCK-OUT equals CLOCK-IN. CLOCK-OUT follows CLOCK-IN until the latter advances the count to 8 (1000), whereupon Q_D clears FF-1, AND-1 stops CLOCK-IN and CLOCK-OUT is low.

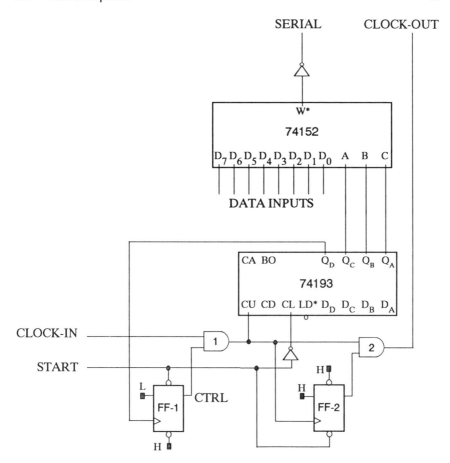

Figure 2.52. Serial conversion circuit application of the 74152 multi-plexer.

It's interesting to consider the sequence SERIAL and CLOCK-OUT undergo without regard to the circuit. Initially, SERIAL equals D_0 and CLOCK-OUT is low. Then, at some arbitrary point in time, CLOCK-OUT goes high and SERIAL becomes D_1. CLOCK-OUT goes high again and SERIAL becomes D_2. The process repeats. When CLOCK-OUT goes high the seventh time, SERIAL becomes D_7. Finally, CLOCK-OUT briefly goes high an eighth time and SERIAL becomes D_0. Because the two outputs could be used by receiving circuitry to reassemble the 8 data bits, the circuit acts as a serial transmitter.

2.10 74373 Latch/buffer and 74244 Buffer

The final digital electronics topic arises from the necessity to connect several outputs together. Often, especially in interfacing circuits, several integrated circuits must at different times put a value on a common control line. Further, devices such as memory and programmable counter/timers must send and receive 8 bit values over a single set of eight lines. This section introduces two integrated circuits which facilitate connecting outputs and establishing bi-directional data flow.

Tri-state outputs are the key to these operations. In addition to the usual low and high, tri-state outputs may be the electrical equivalent of disconnected.

2.10.1 74373 Latch/buffer

The 74373 has eight inputs and eight corresponding outputs as shown in the logic drawing in Figure 2.53.

The chip always contains an 8-bit latch value. When the E (Enable) input goes from high-to-low, the current states of the inputs become the latch value. While E remains low, the value is held. When and while E is high, the latch value equals the input bits and changes with them. So external circuitry must take E low when a value to be stored is on the input connections.

The second 74373 operation is determined by OC* (Output Control). When inactive, the outputs are in a Hi-Z disconnected state. When active,

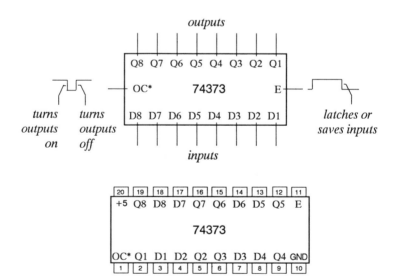

Figure 2.53. Annotated logic drawing and pin assignments for the 74373 latch/buffer with tri-state outputs.

the outputs are the current latch value (which depends on E). So external circuitry must take OC* low to make the latch value available. Output bits from several 74373 can be connected together without difficulty as long as only one OC* is active at a time.

Figure 2.53 also gives the 74373 pin assignments. Note that both the latch and tri-state operations work on all eight bits simultaneously. However, the bits may be thought of either individually or as an 8-bit unit of information.

2.10.2 74244 Tri-state Buffer

Another useful IC is the 74244. While its logic operations are a subset of the 74373's, the chip has electrical characteristics which make it more effective in getting values off and putting values on long noisy cables.

The 74244 has eight inputs and eight corresponding outputs. It has two gate controls G_1* and G_2*. When the gates are inactive, the outputs are all in a disconnected Hi-Z state. When active, each output equals the corresponding input. There is no latch operation. G_1* controls the lower 4 bits and G_2* the upper 4.

Figure 2.54 gives both logic and pin assignment diagrams. The chip is only used in Chapter 7 and always with G_1* and G_2* wired together.

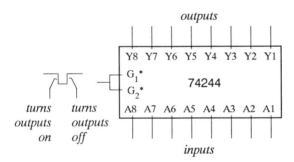

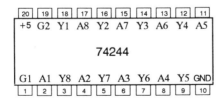

Figure 2.54. Logic drawing and pin assignments for the 74244 line receiver/transmitter.

2.10.3 Interfacing to a Bi-directional Data Bus

Computer expansion slots make bi-directional data lines (which comprise a bus) available to interface circuits. To take advantage of the bus, it's necessary to save values put on the lines by the computer, and, at different times, to put values on the lines which the computer reads. The circuit in Figure 2.55 accomplishes this. Notice in the drawing that the INPUT 74373 has data inputs on the bottom and outputs on the top while the OUTPUT 74373 is inverted.

The INPUT 74373's E control is wired high. This means the latch value always equals the inputs from the interface circuit. For the computer to get a value, OC* must be activated before the computer reads from the lines and deactivated afterward.

The OUTPUT 74373's OC* control is wired active. This means the latch value is always available to the interface circuit. To store a value off the data bus, E must go low while the value is present.

In both cases, additional logic is necessary to generate E and OC* from expansion slot control lines. Chapter 3 presents the necessary circuits for IBM PC/XT/AT and Apple II slots.

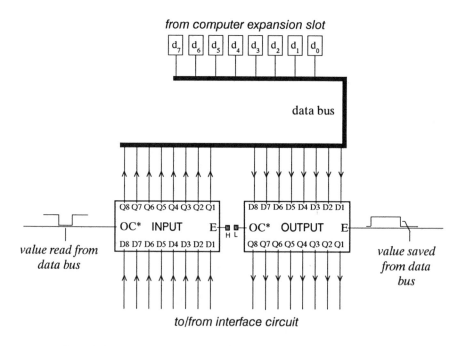

Figure 2.55. 74373's allow interface circuits to send and receive values from an expansion slot's bi-directional data bus.

2.11 Conclusions

The preceding introduction to digital electronics provides the specific background for this book. A number of topics are omitted including arithmetic circuits, 7-segment displays and drivers, and other categories of flip-flops, decoders and multiplexers. Also, Boolean algebra, minimization techniques and other logic families are not covered. As stated earlier, references for more complete treatments are in Appendix A. Even without these topics, the stage is now set to design expansion slot circuits, to interface analog-to-digital converters, memory and programmable counter/times, and to devise comprehensive measurement systems.

Exercises

1. Show how INVERTER and AND gates could be built from NOR's. Use truth tables as needed.

2. Could a control situation analogous to the one in Figure 2.3 be implemented with an OR gate? Describe.

3. Draw a circuit which implements a 4-input NAND gate with 2-input AND's and INVERTERS.

4. Draw a 4-input decoder circuit which produces a low output only when the inputs ABCD are 0000. Sketch circuits which produce low outputs when the inputs are 1111 and 0001.

5. Use a breadboard system to build, test and troubleshoot the following:

   a. The EXCLUSIVE-OR circuit in Figures 2.10 and 2.13.

   b. The decoder circuit in Figure 2.5.

   c. The OR circuit in Figure 2.9.

6. Set up and analyze the situation in Figures 2.18 and 2.19. Specifically, look at the digital clock's output on an oscilloscope. Measure the frequency. Next, use the breadboard system's clock for the high pulse and observe both the breadboard clock and counter signal.

7. Draw a timing diagram which describes the operation of the NAND gate flip-flop in Figure 2.22. Start with SET* and CLEAR* inactive. Include Q2 in the diagram.

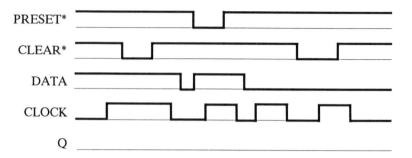

Figure 2.56. Timing diagram for Exercise 9.

8. A flip-flop "type" operation can be produced with two NOR gates if SET and CLEAR are active high. Sketch the circuit and explain its operation.

9. Determine the 7474 output Q for the timing diagram in Figure 2.56.

10. Complete the timing diagram for the circuit in Figure 2.57.

11. Complete the timing diagram for the circuit in Figure 2.58.

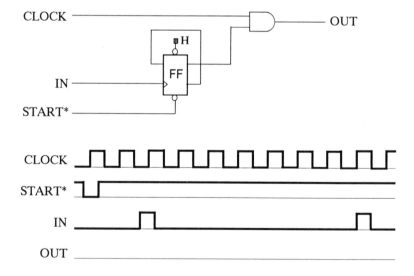

Figure 2.57. Circuit and timing diagram for Exercise 10.

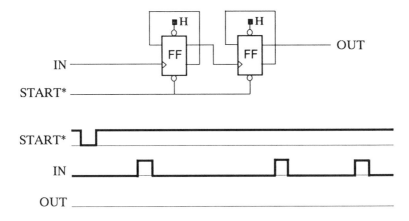

Figure 2.58. Circuit and timing diagram for Exercise 11.

12. In the synchronization circuit in Figure 2.30, what happens when START returns low? Devise a way to observe the timing signal (TS) on an oscilloscope. See if it works.

13. Build and test the LED circuits in Figure 2.32c and d.

14. Build and test the 4-bit dip-switch circuit in Figure 2.34. Use a 7486 EXCLUSIVE-OR chip. Use a 7525 4-input NOR and an INVERTER for the OR.

15. Convert the following hexadecimal numbers to decimal.
    FFFF   7FFF   8000   CDEF   0715

16. Convert the following decimal values to hexadecimal.
    100   1000   10000   25000   1990

17. Convert the following binary numbers to hex.
    1111111111111111   10000000000001100
    0111000000000111   00110111100010110

18. Convert the following hex numbers to binary.
    7FFF   0008   1234   F000

19. Build and test the flip-flop counter in Figure 2.37. Use a breadboard system clock for COUNTS-IN. Observe the clock on oscilloscope channel 1 and successively Q_A, Q_B, Q_C and Q_D on channel 2.

20. Devise a 4-bit binary count-down counter using 7474 flip-flops.

21. Test a 74192 counter as follows: wire CLear low, connect the data inputs to logic switches, connect LoaD* to an active-low pulser, connect COUNT-UP to an active-high pulser, connect the count outputs to LED's and connect CArry to a logic probe. Verify all aspects of counter operation. Then change to the breadboard system's clock set to around 100 KHz for COUNT-UP. Observe the outputs including CArry on an oscilloscope.

22. Build the circuit in Figure 2.42. Test its operation with an oscilloscope. If COUNTS-IN is a 100 KHz clock, is it possible so see LD*'s brief activation?

23. Design a way for a 74193 counter to act as a 74192. (Don't worry about CArry and BOrrow.)

24. Design a circuit which produces a 1.0000 second pulse from a 10.000 KHz symmetrical digital clock.

25. Test a 74138 as follows: wire the gates active, connect logic switches to select inputs C, B and A, and verify with a logic probe that the selected output is low. Successively deactivate G_1, G_{2A}* and G_{2B}* and verify that the gate operates as specified.

26. Expand the application circuit in Section 2.8.2 to utilize six CS*'s and produce six OUT*'s. An additional 74138 chip is required.

27. Answer the following questions about the switch encoding circuit of Section 2.8.3.

    a. What happens if two switches go active simultaneously?

    b. When S_0 is deactivated, what happens if S_3 is active?

28. Sketch a circuit which encodes 16 switches using two 74138's.

29. Sketch a circuit which loads the number of an activated switch into a 3-bit flip-flop memory register (Section 2.4.4).

30. Wire and test the switch encoding circuit in Section 2.8.3. Connect the counter output CBA to LED's. Use the breadboard clock set at 10 KHz and use logic switches for S_3, S_2, S_1 and S_0. Even if the circuit works, go through the troubleshooting tips in Section 2.8.4.

31. Test a 74152 multiplexer as follows. Connect the select inputs (CBA) and D_0 to the breadboard system's logic switches. With CBA = 000, see if W^* is the inverse of D_0. Set CBA to 001, connect D_1 to a logic switch and see if W^* is the inverse of D_1. Repeat for D_2 to D_7.

32. Make a timing diagram for the serial transmitter in Figure 2.52. Include CLOCK-IN, START, CTRL, Q_D, SERIAL and CLOCK-OUT.

33. Design a serial receiver for the transmitter in Figure 2.52. One approach uses a 74138, a 74193, eight flip-flops and basic gates.

34. Figure out how to implement the switch encoder in Figure 2.50 using a 74152, 74193, flip-flop and basic gates. Draw and explain the circuit.

35. Wire and explore a 74373 as follows. Connect OC^* and E to active-low pulsers. Use two or three logic switches for inputs and connect the corresponding outputs to LED's. Latch a value into the chip. Activate OC^* and check the value. When OC^* is inactive, what does a logic probe read for one of the outputs? Finally, wire OC^* active and E high. Do the outputs follow the inputs?

3 Parallel I/O Ports

In the tutorials of Chapters 4-6 and the measurement systems of Chapters 9 and 10, all interactions between a host computer and digital circuits occur through a pair of 8-bit parallel I/O ports. This chapter: 1) gives specifications for the ports; 2) describes in detail their design for IBM PC/XT/AT and Apple II expansion slots; 3) lists commercial vendors who offer suitable ports for these and other computers; 4) outlines test and analysis methods; and 5) presents in detail the three software operations needed throughout the remainder of the book. Readers unfamiliar with the basics of microcomputer architecture are referred to Appendix F which reviews the subject.

3.1 Specifications

The two I/O ports present 32 connections to external circuits as shown in Figure 3.1. Even though the ports are identical, one is designated CONTROL and the other DATA. $CI_7...CI_0$ are the eight Control-In bits. $CO_7...CO_0$ are the Control-Out bits. $DI_7...DI_0$ and $DO_7...DO_0$ are the Data-In and Data-Out bits.

All communication between software running on the host computer and hardware connected to the ports occurs through four program statements. As explained in later sections, each I/O port in an expansion slot is identified by a number or address. Suppose CP designates the Control Port and DP the Data Port. The four program statements, illustrated in BASIC for a PC, are:

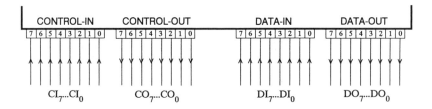

Figure 3.1. All communication with data acquisition circuits occurs through the 32 connections of a pair of 8-bit parallel I/O ports.

1. **OUT CP,value** causes the low 8 bits of [the binary value of] value to appear on the 8 bits of control-out. For example, execution of OUT CP,51 makes $CO_7...CO_0$ = LLHHLLHH (the binary value of 51). Execution of OUT CP,1000 makes $CO_7...CO_0$ = HHHLHLLL, the low byte of the binary value of 1000 = 00000011 11101000.

2. **OUT DP,value** causes the low 8 bits of value to appear on data-out. Value remains on the data-out (or control-out) lines until execution of the next OUT DP,value statement.

3. **variable = INP(CP)** results in variable equaling the 8-bit value presented by external hardware to the control-in bits at the time the statement is executed. For example, if $DI_7...DI_0$ are wired HLLLHHLLL, execution of X = INP(DP) makes X = 152, the binary value of 10011000.

4. **variable = INP(DP)** results in variable equaling the 8-bit value presented to the data-in bits at the time of statement execution.

Section 3.5 explains how software uses the 32 port connections to control and monitor external hardware and to input and output data. The next three sections cover acquiring ports either through construction (Sections 3.2 and 3.3 and Appendices D and E) or purchase (Section 3.4). In each case, the ports circuit is on a board which plugs into one of the host computer's expansion slots. The 32 bits and ground are made available to external data acquisition hardware through appropriate multiconductor cables (supplied by the manufacturer or as described in Appendices D and E).

3.2 Port Design for the IBM PC Expansion Slot

Output on computers with Intel microprocessors (IBM PC/XT/AT and compatibles) requires specifying an 8-bit output value and a 12-bit port number (CP and DP above) which designates the value's destination. Input requires specifying a place to store an 8-bit input value (such as a variable) and a 12-bit port number which designates the value's source. Devices such as the video display, keyboard, disk controller, printer and others have preassigned port numbers which the computer's operating system knows. Numbers for the parallel ports are picked later and must be different from all other I/O devices.

The expansion slots of PC-AT's and compatibles support inputting and outputting 16-bit values. The capability is ignored here.

3.2.1 I/O Sequences

The first step in designing a device for the PC's expansion slot is to figure out which connections are needed. From IBM's *Technical Reference*, the following 23 of 62 connections are required for the parallel ports circuit.

$A_{11}...A_0$ 12 Address lines for the 3 hex digit (12-bit) port number

$d_7...d_0$ 8 data lines for I/O values

control lines   IOR* (Input/Output Read), IOW* (Input/Output Write), AEN (Address ENable)

Figure 3.2 shows the expansion slot and these connections.

After identifying the control, data and address lines, the next step is to understand the timing sequences which occur during input and output operations.

When instructed by software (e.g., execution of X = INP(CP)), the computer's Central Processing Unit (CPU) inputs a value from an I/O de-

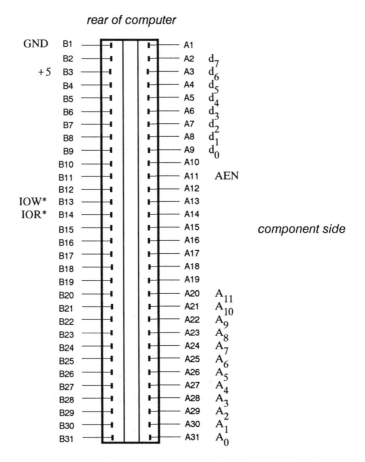

Figure 3.2. Pin assignments for the IBM PC/XT/AT and compatibles expansion slot showing the connections needed for the parallel ports circuit. (PC-AT's and compatibles have an additional 36 pin slot below the one shown.)

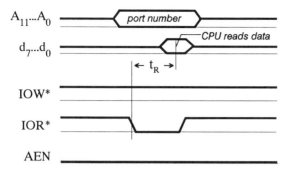

Figure 3.3. Timing pattern for the PC expansion slot input operation. The CPU reads $d_7...d_0$ t_R seconds after making IOR* active.

vice by carrying out the timing sequence in Figure 3.3. Specifically, the CPU puts the desired port number on $A_{11}...A_0$, makes IOR* active, reads the data lines (t_R seconds after activating IOR*), deactivates IOR* and finally removes $A_{11}...A_0$ from the address lines. Whatever I/O device decodes its port number on $A_{11}...A_0$ must respond to the sequence by putting a value on $d_7...d_0$ faster than t_R seconds after IOR* goes active. (t_R is on the order of 500 ns.)

When instructed by software (e.g., execution of OUT CP,51), the CPU outputs a value to an I/O device by implementing the timing sequence in Figure 3.4. Specifically, the CPU puts the desired port number on $A_{11}...A_0$ and the output value on $d_7...d_0$. Then, the CPU briefly activates IOW*. The data value remains at least t_D seconds after IOW* returns inactive. Whatever I/O device decodes its port number on $A_{11}...A_0$ must respond to the sequence by "saving" the value on $d_7...d_0$ within t_D seconds after IOW* returns inactive. (t_D is on the order of 500 ns.)

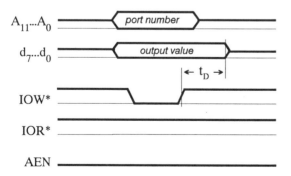

Figure 3.4. Timing pattern for the PC expansion slot output operation. The I/O ports circuit must save the value on $d_7...d_0$ within t_D seconds after IOW* returns inactive.

$A_{11}...A_0$ and $d_7...d_0$ are used for other operations such as storing values in and retrieving values from the computer's memory. However, the CPU activates IOW* only when outputting or writing to an I/O device and only activates IOR* when inputting from or reading an I/O device.

IBM PC expansion slot devices may take control of the computer from the CPU (by activating a control signal not discussed here). When this happens, the control line AEN is active (high) and the ports circuit must not respond to possible I/O sequences. However, the circuit does respond when AEN is low as shown on the timing diagrams in Figures 3.3 and 3.4.

What circuitry is needed for the parallel ports to correctly respond to input and output sequences? Since many I/O operations (usually carried out by the operating system) do not involve the ports, there must be a way for the ports to respond only when $A_{11}...A_0$ contains one of their designated numbers. The next section gives decoder circuitry. Also, during inputs the ports circuit must supply a value and during outputs must store a value. Latch/buffer and controller circuits are given in Sections 3.2.3 and 3.2.4.

3.2.2 Port Number Decoding

IBM's *Technical Reference* states that port numbers hex 300 to 31F are set aside for prototype cards. Therefore, 300 hex is selected for the control port and 301 hex for the data port. (How to choose different numbers is discussed later.) The first goal is a decoder circuit which produces a unique output only when $A_{11}...A_0$ = 300 or 301 hex, or more specifically when $A_{11}...A_0$ = 0011 0000 000X, where X means A_0 may be high or low.

The simplest implementation is to invert A_9 and A_8 and use OR logic as shown in Figure 3.5. Recall that the output of an OR gate is low only if all inputs are low. A_0 is not included in the circuit. The signal DO* (Decoder Out) is active when and only when $A_{11}...A_0$ is 0011 0000 0000 or 0011 000 0001.

In practice, the decoder circuit is complicated because multiple input OR gates are not available. The final circuit, shown in Figure 3.6, consists of 7425 4-input NOR gates and inverters (which convert NOR to OR). Recall that the ports circuit responds to IOW* or IOR* only when AEN is low. This requirement is included in the final decoding circuit by connecting AEN where A_0 is expected. So DO* is active only when $A_{11}...A_0$ = hex 300 or 301 and when AEN is low. As mentioned earlier, A_0 distinguishes port 300 from 301 and is treated separately.

The approach here is to "hardwire" the number the circuit decodes. To respond to a different number, invert additional address bits. If flexibility is required, A_3 and A_2 could be compared with the settings of a 2-bit dip switch (see Section 2.5.3 and Exercise 2).

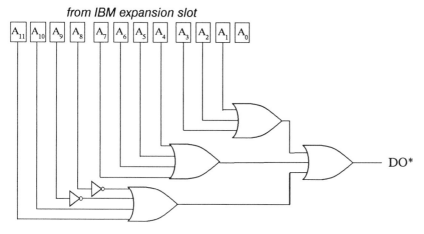

Figure 3.5. Decoder for expansion slot ports circuit whose output DO* is active when and only when $A_{11}...A_0$ is 300 or 301 hex.

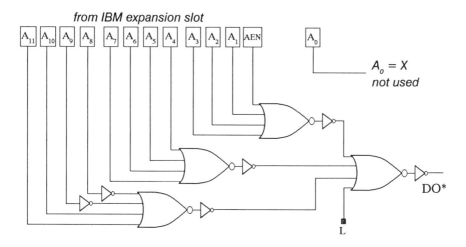

Figure 3.6. Decoder circuit which uses 7425 NOR chips. DO* is active when and only when $A_{11}...A_0$ is 300 or 301 hex and the control AEN is low.

3.2.3 Data Buses and 74373 8-bit Latch/buffers

When the computer executes an output to 300 or 301 hex, the ports circuit must save the value on $d_7...d_0$. And during an input from 300 or 301 hex, the circuit must supply a value to $d_7...d_0$. Figure 3.7 shows how the operations are implemented by four 74373 latch/buffers connected to the expansion slot's $d_7...d_0$.

Recall from Section 2.10 that when a 74373's E (Enable) control is high, each output (Q_i) equals the corresponding input (D_i). When E goes low, the current input value is stored. Put another way, when E is low, $Q_i = D_{io}$ where D_{io} is the value at the input the last time E went low. Recall also that the output pins are in a Hi-Z disconnected state unless OC* (Output Control) is active.

In Figure 3.7, the E's of the control and data-in 74373's are wired high. Therefore, whenever either OC-C* or OC-D* is active, the current value supplied by external hardware is put on $d_7...d_0$. Otherwise, both chips' outputs are in a Hi-Z state and do not interfere with data bus operations.

During both control and data-in, the decode output DO* is active. Figure 3.8 reproduces the expansion slot input timing pattern from Figure 3.3 except $A_{11}...A_1$ and AEN are replaced by DO* and A_0 is shown separately. As can be seen from the figure, if OC-C* follows IOR* when DO* is active and $A_0 = 0$, then the control-in 74373 supplies a value to $d_7...d_0$ which the CPU properly reads. Also, if OC-D* follows IOR* when DO* is active and $A_0 = 1$, then the data-in 74373 supplies a proper value to $d_7...d_0$. (Note the gate delays between IOW* and OC-C* and OC-D*. They are discussed in Section 3.2.5.)

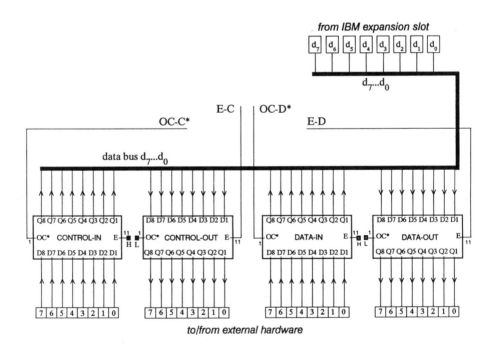

Figure 3.7. Four 74373 latch/buffer IC's transfer values to and from $d_7...d_0$ under the control of OC-C*, OC-D*, E-C and E-D.

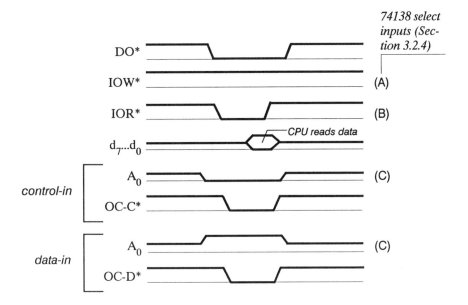

Figure 3.8. Control and data-in timing patterns. While OC-C* is active, the control-in 74373 puts a value on $d_7...d_0$. While OC-D* is active, the data-in 74373 puts a value on $d_7...d_0$.

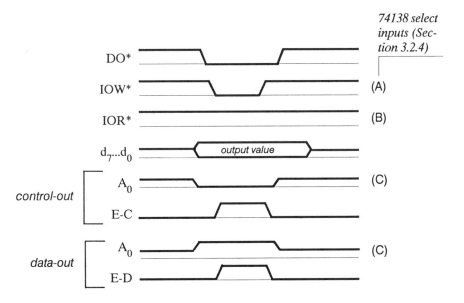

Figure 3.9. Control and data-out timing patterns. When E-C goes low, the control-out 74373 saves the current value on $d_7...d_0$. When E-D goes low, the data-out 74373 save the value.

The OC*'s of the control and data-out 74373's are wired low. This means external hardware has continual access to the values stored by the chips, which are the values that were on $d_7...d_0$ the last time E-C and E-D went low.

During both control and data-out, DO* is active. Figure 3.9 reproduces the expansion slot output timing pattern from Figure 3.4 with $A_{11}...A_1$ and AEN replaced by DO* and with A_0 shown separately. As can be seen in the figure, if E-C follows the inverse of IOW* when DO* is active and $A_0 = 0$, then the control-out 74373 properly stores the output value off $d_7...d_0$. If E-D follows the inverse of IOW* when DO* is active and $A_0 = 0$, then the data-out 74373 stores a proper output value. (The gate delays between IOW* and E-C and E-D are discussed in Section 3.2.5.)

3.2.4 Parallel Port Controller

The final problem is how to produce OC-C*, OC-D*, E-C and E-D. The job could be done with basic gates. However, greater flexibility is attained by a 74138 decoder chip.

Recall the 74138 decoder/demultiplexer from Section 2.8. The chip's logic drawing is reproduced in Figure 3.10. If G_1 and G_{2A}^* and G_{2B}^* are all active, the Y_i output selected by the current value of $i = CBA$ goes and remains low as long as the gates are active and the output selected. Otherwise, the selected and all other Y's are high.

How can the 74138 produce the four 74373 controls? Suppose the decoder output (DO*) is connected to both G_{2A}^* and G_{2B}^* and G_1 is wired high. This means the selected output goes low only when DO* is active (only when $A_{11}...A_0$ = 300 or 301 hex and AEN is low).

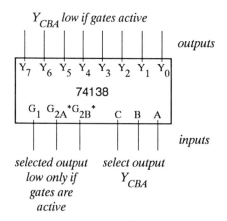

Figure 3.10. Logic drawing of the 74138 decoder/demultiplexer.

Next suppose the 74138 select inputs are:

C	B	A
A_0	IOR*	IOW*

As shown in Figure 3.8, OC-C* is low only when CBA = LLH = 1. Hence output Y_1 is exactly what OC-C* needs to be. Also, OC-D* is low only when CBA = HLH = 5. Hence output Y_5 = OC-D*. As shown in Figure 3.9, E-C is high only when CBA = LHL = 2. Hence, Y_2 inverted = E-C. Also E-D is high only when CBA = HHL = 6. Hence Y_6 inverted = E-D. These results are summarized in the table below.

C	B	A	selected	
A_0	IOR*	IOW*	output	signal
L	L	H	Y_1	OC-C*
H	L	H	Y_5	OC-D*
L	H	L	Y_2	E-C
H	H	L	Y_6	E-D

The complete I/O ports circuit is given in Figure 3.11. A parts list and construction instructions are in Appendix D. Test and application software is in Section 3.5. Troubleshooting is also discussed in Appendix D.

3.2.5 Critical Timing and Gate Delays

Now that the PC ports circuit is determined, it's important to think over the timing and look for possible "glitches." Two factors are involved. The first is the expansion slot timing sequences given in Figures 3.3 and 3.4. The second is gate delays in the ports circuit.

Consider the input sequence in Figure 3.8. The critical question is: does the circuit put a value on $d_7...d_0$ before the CPU reads the lines? A 74LS373 integrated circuit responds to its OC* going active by putting a value on its output pins (Q's) within 36 ns. The delay between IOR* and either OC* going active is introduced by the 74138. For an LS chip, the maximum is 48 ns. So there is up to 84 ns between IOR* going active and a valid value on $d_7...d_0$. Therefore, if the CPU reads $d_7...d_0$ more than 84 ns after IOR* goes active, the circuit works. For the IBM expansion slot, this time t_R, shown in Figure 3.3, is on the order of 500 ns..

Consider the output sequence in Figure 3.9. According to specifications for a 74LS373 IC, values must remain at the data pins 10 ns after the enable control goes low. In the ports circuit, the E's follow IOW* but are delayed up to 60 ns by the 74138 and inverters. Therefore, the value on $d_7...d_0$ must remain at least 70 ns after IOW* returns inactive. For the IBM expansion slot, the available time t_D, shown in Figure 3.4, is on the order of 500 ns.

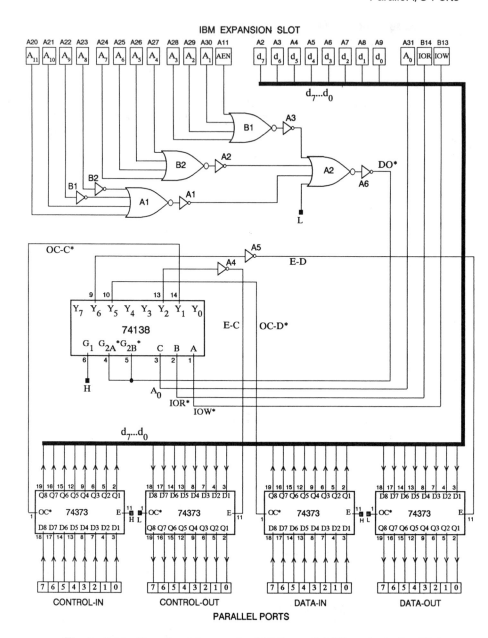

Figure 3.11. The circuit for a pair of 8-bit parallel I/O ports for the IBM PC and compatibles' expansion slot. Pin numbers are given for major IC's. NOR and INVERTER gate labels are used in Appendix D where port construction is described.

As computer speeds increase (from the original PC to the XT, AT, fast compatibles and to 80386 systems), expansion slot timing is maintained so circuits which work on an original PC also work on AT's and fast compatibles. However, if faster slots come along, the times have to be reevaluated and it may become necessary to use faster integrated circuits.

3.3 Port Design for the Apple II Expansion Slot

Input/output on Apple II and compatible computers requires peeking 8-bit values from and poking 8-bit values to specific memory addresses.

Apple II's have 8 expansion slots numbered 0 to 7. All slots are given access to 16 address and 8 data lines as well as a variety of controls. Each slot is allocated 16 bytes of the computer's memory space. When a value is peeked from or poked to an address in the range allocated for a slot, a control line DS* (Device Select) unique to that slot is activated. The address in the range is put on the $A_{15}...A_0$ and the I/O value is communicated on $d_7...d_0$. The R/W* (Read/Write) control distinguishes read (peek or input) from write (poke or output).

A second, independent range of 256 memory bytes is allocated to slots 0 to 7. Also, each slot has another unique control (I/O Select) which indicates a peek or poke in the slot's range. This second, larger memory allocation is not needed for the parallel ports.

The table below gives the address ranges which when peeked or poked cause activation of a particular slot's DS*.

slot number	address range hex	decimal
0	C080-C08F	49280-49295
1	C090-C09F	49296-49311
2	C0A0-C0AF	49312-49327
3	C0B0-C0BF	49328-49343
4	C0C0-C0CF	49344-49359
5	C0D0-C0DF	49360-49375
6	C0E0-C0EF	49376-49391
7	C0F0-C0FF	49392-49407

So during execution of a peek or poke to 49328, the DS* of slot 3 (only) is activated. A peek or poke to 49331 also causes activation of slot 3's DS*. On the other hand, a peek or poke to 49344 causes activation of slot 4's DS*.

3.3.1 Apple II I/O Sequences

The goal is to design a pair of 8-bit parallel I/O ports which work in the Apple II expansion slot. The first step is to decide the slot dependent addresses for CI (Control-In), CO (Control-Out), DI (Data-In) and DO (Data-Out). The following scheme is adopted.

Control-in	CI	1st address in the range of the slot occupied by the ports circuit
Control-out	$CO = CI + 1$	2nd address in the range
Data-in	$DI = CI + 2$	3rd address
Data-out	$DO = CI + 3$	4th address

If the ports circuit is in slot 3, the addresses in decimal, hex and binary (low two bits) are:

	address		
constant	decimal	hex	$A_1 A_0$
CI	49328	C0B0	0 0
CO	49329	C0B1	0 1
DI	49330	C0B2	1 0
DO	49331	C0B3	1 1

Only the low two bits are required to distinguish the four addresses no matter which slot is occupied. The remaining 12 memory locations of the occupied slot are not used unless more than two ports are required (see Exercise 4).

From the Apple IIe *Reference Manual* and other references (all given in Appendix A), the following 12 of 50 connections are required for the ports circuit.

A_1, A_0	The low 2 of 16 address lines
$d_7 ... d_0$	8 lines for I/O values
DS*	The **D**evice **S**elect of whatever slot the circuit occupies
R/W*	**R**ead/**W**rite control (high for read, low for write)

Figure 3.12 shows the expansion slot and these connections. (+5 and ground are also shown.)

After assigning port addresses and identifying the control, data and addresses lines, the next step is to understand the timing sequences which occur during peek and poke operations.

The diagrams below were determined from the Apple IIe Reference Manual, other references in Appendix A, and measurements with an oscilloscope.

Input or peek from the expansion slot occurs as shown in Figure 3.13. The CPU puts the peek address on $A_{15} ... A_0$. If the address is in the range allocated for the expansion slot occupied by the ports circuit, DS* is activated. R/W* remains high. t_R seconds after DS* activation, the CPU reads $d_7 ... d_0$ for the input value. On 1989 model Apple IIe's, t_R is on the order of 500 ns.

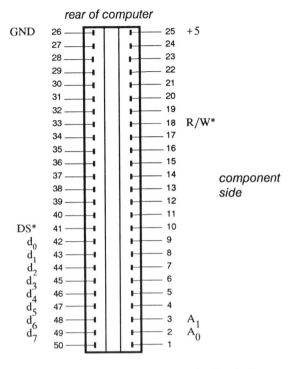

Figure 3.12. Pin assignments for the Apple II expansion slot showing the connections needed for the parallel ports circuit.

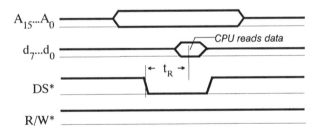

Figure 3.13. Timing pattern for the Apple II expansion slot input/read/ peek operation. The CPU reads $d_7...d_0$ t_R seconds after DS* goes active.

Output or poke from the computer occurs with the timing sequence given in Figure 3.14. The CPU puts the poke address on $A_{15}...A_0$ and the output value on $d_7...d_0$. R/W* is taken low. Then, DS* is activated (only if the poke address is in the range allocated for the slot occupied by the ports circuit). The output value remains t_D seconds after DS* returns inactive. On 1989 model Apple IIe's, t_D is on the order of 150 ns.

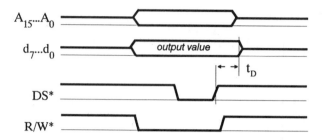

Figure 3.14. Timing pattern for the Apple II expansion slot output/write/ poke operation. The I/O ports circuit must save the value on $d_7...d_0$ within t_D seconds after DS* returns inactive.

On earlier model Apple II's, during a poke sequence, DS goes active twice. The first time is earlier in the sequence shown in Figure 3.14 while R/ W* is high. The ports circuit described here also works in this case.*

In order to respond to the sequences, the circuit must determine when a port is addressed and save the value on $d_7...d_0$ during outputs and put a value on $d_7...d_0$ during inputs. This is done with the latch/buffers and controller discussed below.

3.3.2 Data Buses and 74373 8-bit Latch/buffers

Figure 3.15 shows how peek and poke operations are implemented by four 74373 latch/buffers connected to the expansion slots $d_7...d_0$.

Recall from Section 2.9 that when a 74373's E (Enable) control is high, each output (Q_i) equals the corresponding input (D_i). When E goes low, the current input value is stored. Put another way, when E is low, $Q_i = D_{io}$ where D_{io} is the value at the input the last time E went low. Recall also that the output pins are in a Hi-Z disconnected state unless OC* (Output Control) is active.

In Figure 3.15, the E's of the control and data-in 74373's are wired high. Therefore, whenever either OC-C* or OC-D* is active, the current value supplied by external hardware is put on $d_7...d_0$. Otherwise, both chips' outputs are in a Hi-Z state and do not interfere with data bus operations.

The input timing sequence in Figure 3.13 is reproduced in Figure 3.16 with A_1 and A_0 given separately. As can be seen in the figure, if OC-C* follows DS* when R/W* is high and A_1A_0 = 00, then the control-in 74373 supplies a value to $d_7...d_0$ which the CPU properly reads. If OC-D* follows DS* when R/W* is high and A_1A_0 = 10, then the data-in 74373 supplies a proper value to $d_7...d_0$. (Recall that the low 2 bits of control-in's address are 00, and data-in's low 2 bits are 10.)

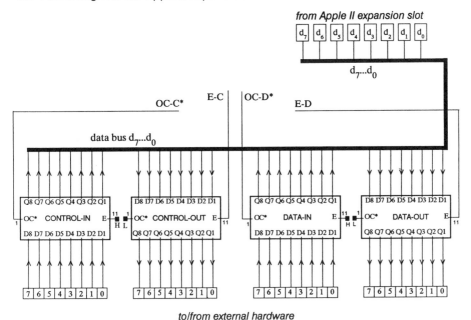

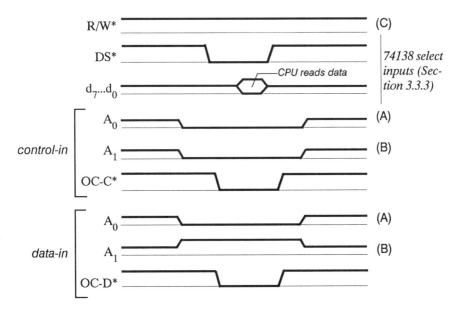

Figure 3.15. Four 74373 latch/buffer IC's transfer values to and from $d_7...d_0$ under the control of OC-C*, OC-D*, E-C and E-D.

Figure 3.16. Control and data-in timing patterns. While OC-C* is active, the control-in 74373 puts a value on $d_7...d_0$. While OC-D* is active, the data-in 74373 puts a value on $d_7...d_0$.

The OC*'s of the control and data-out 74373's are wired low. This means external hardware has continual access to the values stored by the chips, which are the values that were on $d_7...d_0$ the last time E-C and E-D went low.

The output timing sequence in Figure 3.14 is reproduced in Figure 3.17 with A_1 and A_0 given separately. As can be seen in the figure, if E-C follows the inverse of DS* when R/W* is low and $A_1A_0 = 01$, then the control-out 74373 properly stores the output value off $d_7...d_0$. If E-D follows the inverse of DS* when R/W* is active and $A_1A_0 = 11$, then the data-out 74373 stores a proper output value. (Recall that the low two bits of control-out's address are 01, and data-out's low 2 bits are 11.)

3.3.3 Apple Parallel Port Controller

The final problem is to produce OC-C*, OC-D*, E-C and E-D. The job could be done with basic gates. However, greater flexibility is attained by a 74138 decoder chip.

Recall the 74138 decoder/demultiplexer from Section 2.8. The chip's logic drawing is reproduced in Figure 3.18. If gates G_1 and G_{2A}^* and G_{2B}^* are active, the Y_i output selected by the current value of $i = CBA$ goes low and re-

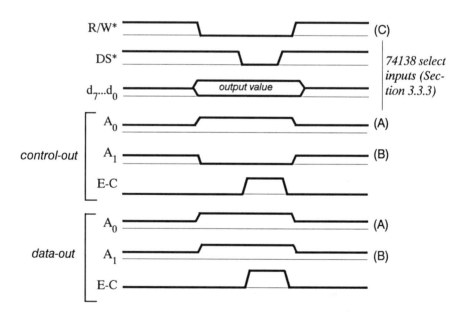

Figure 3.17. Control and data-out timing patterns. When E-C goes low, the control-out 74373 saves the current value on $d_7...d_0$. When E-D goes low, the data-out 74373 save the value.

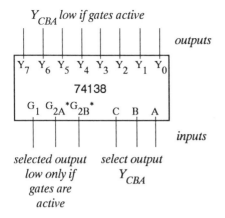

Y_{CBA} *low if gates active*

outputs

| Y_7 | Y_6 | Y_5 | Y_4 | Y_3 | Y_2 | Y_1 | Y_0 |

74138

G_1 $G_{2A}{}^*G_{2B}{}^*$ C   B   A

inputs

selected output   *select output*
low only if Y_{CBA}
gates are
active

Figure 3.18. Logic drawing of the 74138 decoder/demultiplexer.

mains low as long as the gates are active and the output selected. Otherwise, the selected and all other Y's are high.

How can the 74138 produce the 74373 controls? Suppose Device Select (DS*) is connected to both $G_{2A}{}^*$ and $G_{2B}{}^*$ and G_1 is wired high. This means the selected output can go low only when DS* is active. Suppose the 74138 select inputs are:

C	B	A
R/W*	A_1	A_0

As shown in Figure 3.16, OC-C* is low only when CBA = HLL = 4. Hence, output Y_4 is exactly what OC-C* needs to be. Also, OC-D* is low only when CBA = HHL = 6. Hence, output Y_6 = OC-D*. As shown in Figure 3.17, E-C is high only when CBA = LLH = 1. Hence, Y_1 inverted = E-C. Also E-D is high only when CBA = LHH = 3. Hence Y_3 inverted = E-D. These results are summarized in the table below.

C	B	A	selected	
R/W*	A_1	A_0	output	signal
H	L	L	Y_4	OC-C*
H	H	L	Y_6	OC-D*
L	L	H	Y_1	E-C
L	H	H	Y_3	E-D

The complete I/O ports circuit is given in Figure 3.19.

A parts list and construction instructions for the ports are given in Appendix E. Test and application software is in Section 3.5. Troubleshooting is also discussed in Appendix E.

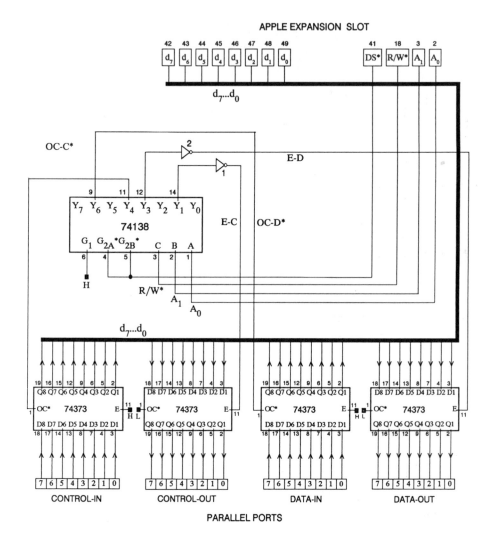

Figure 3.19. The complete circuit for a pair of 8-bit parallel I/O ports for the Apple II expansion slot.

3.3.4 Critical Timing and Gate Delays

Now that the Apple ports circuit is determined, it's important to think over the timing and look for possible "glitches." Two factors are involved. The first is the expansion slot timing sequences given in Figures 3.13 and 3.14. The second is gate delays introduced by the ports circuit.

Consider the input sequence in Figure 3.16. The critical question is: does the circuit put a value on $d_7...d_0$ before the CPU reads the lines? A

74LS373 integrated circuit responds to its OC* going active by putting a value on its output pins (Q's) within 36 ns. The delay between DS* and either OC* going active is introduced by the 74138. For an LS chip, the maximum is 48 ns. So there is up to 84 ns between DS* going active and a valid value on $d_7...d_0$. Therefore, if the CPU reads $d_7...d_0$ more than 84 ns after DS* goes active, the circuit works. For a 1989 Apple IIe expansion slot, the time t_R, shown in Figure 3.13, is on the order of 500 ns..

Consider the output sequence in Figure 3.17. According to specifications for a 74LS373 IC, values must remain at the data pins 10 ns after the enable control goes low. In the ports circuit, the E's follow DS* but are delayed up to 60 ns by the 74138 and inverters. Therefore, the value on $d_7...d_0$ must remain at least 70 ns after DS* returns inactive. For an Apple IIe expansion slot, the available time t_D, shown in Figure 3.14, is on the order of 150 ns.

3.4 Commercial Parallel I/O Ports

The analog-to-digital converter, memory and programmable counter/timer described in the next three chapters and the measurement systems in the last four require two 8-bit output ports and two 8-bit input ports. All example software works through control and data-in and control and data-out. An alternative to constructing ports for PC's or Apple II's is to purchase them. Further, Macintosh and IBM PS/2 users must acquire suitable ports before taking advantage of the following section and chapters.

A number of manufacturers make I/O boards for computer expansion slots. While new products are continually introduced and old ones discontinued, boards with 32 bits are likely to always be available for popular computers. During 1989, a board suitable for PC/XT/AT and compatibles can be purchased from Anasco for $200, and an Apple II board from Applied Engineering for $90. More expensive and complex systems are available for Macintosh and IBM PS/2 computers. The purpose of this section is to specify what's required for commercial ports and list companies which according to their catalogues produce suitable systems.

3.4.1 Specifications and Pitfalls

In selecting a commercial system, the first thing to look for is the number of I/O bits. There must be at least 32 which can be divided into four sets of 8 bits. Each set must be individually accessible to software as described in Section 3.1. A common characteristic is that I/O bits may be programmed for input or output. This requires additional setup software. For PC and PS/2 computers, it's necessary to determine port numbers. For Apple II computers, the peek and poke addresses must be known. Boards for Macintosh computers come with software drivers and I/O is accomplished with system type calls.

An essential requirement is that 8-bit output values remain latched from one output until the next.

Commercial boards sometimes perform more than just I/O operations. Unneeded functions such as counting, timing and interrupt control should be ignored. If at all possible avoid boards which also support analog conversions. Parallel printer cards are not acceptable.

3.4.2 IBM PC/XT/AT and Compatible Computers

By far the greatest variety of expansion slot I/O boards are available for IBM PC/XT/AT and compatible computers. Here are eight vendors.

Anasco Corp.
42A Cherry Hill Drive
Danvers, MA 01923

Intelligent I/O
1141 West Grant Rd. #131
Tucson, AZ 85705

ICS Electronics Corp.
2185 Old Oakland Road
San Jose, CA 95131

Metrabyte
440 Myles Standish Blvd.
Taunton, MA 02780

Burr-Brown Corp.
P.O. Box 5100
Carrollton, TX 75011

Perx
1730 South Amphlett Blvd.
Suite 222
San Mateo, CA 94402

Rapid Systems, Inc.
433 N. 34th Street
Seattle, WA 98103

Action Instruments, INC.
8601 Aero Drive
San Diego, CA 92123

3.4.3 IBM PS/2 Computers

Boards for the IBM PS/2 micro channel expansion slot are available from a growing number of companies including these two.

ICS Electronics Corp.
2185 Old Oakland Road
San Jose, CA 95131

Metrabyte
440 Myles Standish Blvd.
Taunton, MA 02780

3.4.4 Apple II Computers

Applied Engineering
P.O. Box 5100
Carrollton, TX 75011

Metrabyte
440 Myles Standish Blvd.
Taunton, MA 02780

3.4.5 Macintosh and Macintosh II Computers

The Macintosh line of computers does not have an expansion slot. The only company found which may supply the equivalent of parallel ports is:

G.W. Instruments
P.O. Box 2145
264 Msgr O'Brien Hwy
Cambridge, MA 02141

The Macintosh II line has an expansion slot, and a growing number of companies offer I/O boards.

G.W. Instruments   ICS Electronics Corp
P.O. Box 2145   2185 Old Oakland Road
264 Msgr O'Brien Hwy   San Jose, CA 95131
Cambridge, MA 02141

National Instruments
12109 Technology Blvd.
Austin, TX 78727

3.5 Software

All interactions between software running on a host computer and measurement circuits connected to the parallel ports occur through four program statements. For the PC and Apple ports described in Sections 3.2 and 3.3, the statements are given below in Microsoft QuickBASIC and Applesoft BASIC. Equivalent statements must be determined for commercial ports and for different languages.

operation	PC QuickBASIC	Apple II Applesoft
control-out	`OUT CP,value`	`POKE CO,value`
data-out	`OUT DP,value`	`POKE DO,value`
control-in	`variable = INP(CP)`	`variable = PEEK(CI)`
data-in	`variable = INP(DP)`	`variable = PEEK(DI)`
constants	`CP = &H300`	`CI = 49328`
	`DP = &H301`	`CO = CI + 1`
		`DI = CI + 2`
		`DO = CI + 3`

The constants apply to PC ports wired to decode 300 and 301 hex and Apple II ports in slot 3.

All software in this book is written in Microsoft QuickBASIC. Equivalent programs in Applesoft BASIC and other languages are straightforward. The software focuses on interactions with hardware and is kept as simple as possible. No graphics, file I/O or other operations are included. However, because of the popularity of PC's and the low cost of QuickBASIC, several features of the language are exploited. For example, in assigning values to constants in the table above, hexadecimal values are designated by the prefix "&H".

Before proceeding to the device tutorials in Chapters 4-6 and the data acquisition systems in Chapters 7-10, it's important to have a clear picture of what happens when the four statements are executed. The best way to thoroughly understand the operations is to acquire parallel ports, determine the four language statements and appropriate constants (if different from above), enter the programs in the remainder of this chapter, and observe what happens when they run. A breadboard system, logic probe and oscilloscope are also needed. In addition to demonstrating the parallel ports, the programs in Sections 3.5.1 and 3.5.2 may be used to test ports hardware. Programs in Sections 3.5.3, 3.5.4 and 3.5.5 illustrate the three software operations needed for all measurement circuits in this book. And Section 3.5.6 illustrates troubleshooting circuits connected to the ports.

3.5.1 Test Control-out and Data-out

The program below asks for the port number in decimal (768 decimal = 300 hex and 769 = 301). It then asks for a value between 0 and 255. As soon as a number is entered, its binary value is output and remains on the lines until the next output.

```
'       Program TESTOUT
        DEFINT A-Z
        INPUT "Enter Port Number (768 or 769) "; PORT
TESTO: INPUT "Enter 0 to 255 or > 255 to END"; OVAL
        IF OVAL > 255 THEN GOTO FINI
        OUT PORT,OVAL
        GOTO TESTO
FINI:  END
```

The eight lines of control-out or data-out can be tested with a logic probe (which distinguishes ground from disconnected). A reasonable procedure is to first output the decimal value "0" and test all bits with the probe. Then output "1" and test all bits. Continue by outputting 2, 4, 8, 16, 32, 64 and 128. In each case, one bit should be high and the rest low. End and restart the program to test the other port.

In QuickBASIC and most languages, constants and variables default to real formats. Since I/O operations involve 8-bit binary values, real numbers must be converted to binary before outputs, and, after inputs, binary values must be converted to real. To avoid these time-consuming manipulations, a good practice is to make all variables and constants integer. This is done in QuickBASIC with the DEFINT A-Z statement. When real or other data

types are needed, the declaration may be overridden by appending a designated symbol ("!" for single precision reals) to variable and constant names.

3.5.2 Test Control-in and Data-in

The program below asks for a port number (768 decimal = 300 hex or 769 =
301). It then inputs from the port, displays the value in decimal, and waits for
an < ENTER > to repeat or an END to terminate.

```
'           Program TESTIN
            DEFINT A-Z
            INPUT "Enter Port Number (768 or 769) "; PORT
TESTI: INVAL = INP(PORT)
            PRINT INVAL
            INPUT "Type <enter> to repeat, else END "; ANS$
            IF ANS$ = "END" OR ANS$ = "end" THEN GOTO FINI
            GOTO TESTI
FINI:  END
```

The input value depends on the wires connected to the control-in or
data-in lines. An approach is to wire bit zero low (with the other bits disconnected). Start the program and see if the input value is correct. Then, wire
the next higher bit low, and repeat. Recall from Section 2.5.1 that a disconnected TTL input acts the same as if it were wired high. Hence the input
values should follow this pattern:

bit wired low	binary	input value hex	decimal
0	1111 1110	FE	254
1	1111 1101	FD	253
2	1111 1011	FB	251
3	1111 0111	F7	247
4	1110 1111	EF	239
5	1101 1111	DF	223
6	1011 1111	BF	191
7	0111 1111	7F	127

3.5.3 Generation of Control Signals

Only three software operations are needed for all measurement circuits.
They are: generating control signals, monitoring circuit status, and inputting
and outputting data. This and the next two sections cover the operations.

Often circuits connected to the parallel ports require software generated control signals. A signal could be as simple as a brief high pulse to clear
a 74193 counter or a brief low pulse to load data into the counter. On the
other hand, devices such as programmable interval timers require several
control signals which carry out timing sequences such as the one in Figure

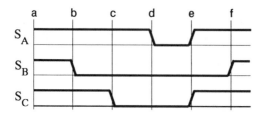

Figure 3.20. Timing pattern generated by software through a sequence of outputs to the control port.

3.20. A method to generate the sequences is given here. Once understood, it can be applied to virtually any timing pattern (with up to eight controls).

The goal is for software to produce three digital signals which follow the sequence in the figure. S_A, S_B and S_C originate at the control port and their states are determined by the most recent output to the port. The first decision is which bit to use for which signal. In elaborate cases such as the one in Chapter 7, it's important to make good choices. But, to begin with, any choice will do. Suppose the signal S_A is connected to control-out bit 2, S_B to bit 3 and S_C to bit 4. The table below shows the choices in the context of all eight control-out bits. "X" means no connection.

CO_7	CO_6	CO_5	CO_4	CO_3	CO_2	CO_1	CO_0
X	X	X	S_C	S_B	S_A	X	X

To produce the sequence in the figure, values are successively output with bits 3, 4 and 5 chosen to make the controls follow the pattern.

Before determining the values, consider the first and last states of each control. In general, these states cause no action in a circuit and are henceforth called normal or inactive. Further, they are usually high because most control signals are inactive high (as explained in Section 2.5.1).

The table below shows how to determine the normal state and other output values. Included in the table is the point in Figure 3.20 which corresponds to each output, the hex value, and a constant name for use in software. A rule is adopted that when a bit doesn't matter (such as the unconnected control-out bits), a high or one is output.

point	CO_7 X	CO_6 X	CO_5 X	CO_4 S_C	CO_3 S_B	CO_2 S_A	CO_1 X	CO_0 X	hex value	constant name
a	1	1	1	1	1	1	1	1	FF	NORM
b	1	1	1	1	0	1	1	1	F7	STEP1
c	1	1	1	0	0	1	1	1	E7	STEP2
d	1	1	1	0	0	0	1	1	E3	STEP3
e	1	1	1	1	0	1	1	1	F7	STEP1
f	1	1	1	1	1	1	1	1	FF	NORM

What happens when these values are successively sent to control-out? The first output makes all bits high and establishes point a in the figure. The second output makes S_B only go low (point b). The third output makes S_C and S_B low (c). The next output makes all three controls low (d). The next returns S_C and S_A high (e). And the final output restores the normal, pre-sequence states (f).

The following program repeatedly generates the pattern until a key is typed.

```
'          Program MAKECTRL
           DEFINT A-Z
           CP = &H300
           NORM = &HFF              ' 1111 1111 binary
           STEP1 = &HF7             ' 1111 0111 binary
           STEP2 = &HE7             ' 1110 0111 binary
           STEP3 = &HE3             ' 1110 0011 binary
REPT:      OUT CP,NORM              ' a
           OUT CP,STEP1             ' b
           OUT CP,STEP2             ' c
           OUT CP,STEP3             ' d
           OUT CP,STEP1             ' e
           OUT CP,NORM              ' f
           IF INKEY$ = "" THEN GOTO REPT
FINI:      END
```

What actually happens when the program runs can be observed by connecting control-out bits 3 and 2 and then 3 and 4 to an oscilloscope and displaying the pattern. The reason for putting the output sequence in a loop is to establish the periodicity required by the scope.

What is the time scale in Figure 3.20? How long do the steps last? The answer depends on the rate of program execution, which in turn depends on the language, CPU speed and other factors. The example program gives the maximum speed. If necessary, the steps can be made longer by repeating the same statement two or more times.

In much of the rest of this book, it's important to know the time between outputs for the particular computer and language being used. In the example case, this is the time between steps d and e which is how long S_A is low. The time should be measured on an oscilloscope and the value recorded for future reference.

For a 12.5 MHz PC-AT compatible using the program above in a stand-alone compiled form, the time is 1.1 μsec.

3.5.4 Polling Hardware Status

The second operation needed for measurement circuits is a way to coordinate hardware events with software. For example, software must have a way to know when data is ready from an ADC. In the parallel ports methodology, coordination is implemented by a polling technique. One or several signals are connected to the control-in bits. The program continually inputs from the port, analyzes the bit pattern, and repeats inputting until a specific condition is met. Only then does the program proceed.

To illustrate the technique and demonstrate the range of possibilities, suppose the program sends (via control-out) a start pulse to a data acquisition circuit. Suppose the circuit responds by generating the Monitor Signal (MS) shown in Figure 3.21. Suppose MS is connected to control-in bit 3. Finally, suppose software must wait until MS first goes high and then back low. While waiting, all the other control-in bits may be high or low.

After sending the start signal, the operation

        INVAL = INP(CP)

yields

CI_7	CI_6	CI_5	CI_4	CI_3	CI_2	CI_1	CI_0	
X	X	X	X	MS	X	X	X	INVAL

where X means the bit may be high or low and CP equals 300 hex.

The unneeded bits are masked by a software AND statement as follows:

        INVAL = INVAL AND MASK

The operation AND's corresponding bits of INVAL and MASK and assigns the results to INVAL as shown in the table below.

X	X	X	X	MS	X	X	X	INVAL
0	0	0	0	1	0	0	0	MASK
0	0	0	0	MS	0	0	0	AND result

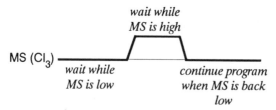

Figure 3.21. Signal produced by a measurement circuit which software monitors through control-in.

The value of MASK depends on the control-in bit or bits of interest. For CI_3, MASK = 8. After AND, if INVAL = 8 = MASK, MS was high when the input occurred. If INVAL = 0, MS was low.

One of the major differences between Applesoft BASIC and QuickBASIC is that the former does not support the logical AND operation. Therefore, slow division and integer statements must be used to determine if a particular bit is high or low. For the case above, it's necessary to perform the operation INVAL = INT(INVAL/2) three times which shifts MS to the low bit. Then, if INT(INVAL/2) = INVAL/2, the final value is even and MS is low. Otherwise it is high. Of course, 6502 assembly language supports the AND operation which makes polling again straightforward.

The program below first Waits While MS (CI_3) is Low (WWL) and then Waits While MS is High. Execution continues beyond WWH only when MS is back low.

```
'       Program POLLMS
        DEFINT A-Z
        CP = &H300
        MASK = &H08
           ...
        other introductory operations
           ...
        statements which send the start signal
WWL:    IF (INP(CP) AND MASK) = 0 THEN GOTO WWL
WWH:    IF (INP(CP) AND MASK) = MASK THEN GOTO WWH
        program statements which follow waits
```

Combining the INPut, AND, IF and GOTO operations in a single BASIC statement makes the loops clear and concise.

The polling process can be visualized by connecting CI_3 to a breadboard system's active-high pulser, and then entering and running the slightly modified program below.

```
'       Program TESTPOLL
        DEFINT A-Z
        CP = &H300
        MASK = &H08
        CLS
        PRINT "Waiting for MS (CI3) high"
WWL:    IF (INP(CP) AND MASK) = 0 THEN GOTO WWL
        PRINT "MS is now high.  Waiting for low"
WWH:    IF (INP(CP) AND MASK) = MASK THEN GOTO WWH
        PRINT "MS is back low"
        END
```

There is a major danger with the polling technique. In Figure 3.21, what if no INP occurs while MS is high? This could happen if the hardware sequence is very fast or the computer very slow. In either case, the program hangs-up in the WWH loop because the IF condition is always satisfied. When there is even the remotest possibility of missing an action (such as the high state here), flip-flops must be added which hold a state until the software responds. The technique is illustrated in the next chapter.

It's important to know how long hardware actions must last to ensure at least one INP occurs during the action. The length of a typical polling loop can be determined by putting a control output at the beginning and end of the loop. For instance,

```
        Program POLLTIME
        DEFINT A-Z
        CP = &H300
        MASK = &H08
WWL:    OUT CP,0
        IF INP(CP) AND MASK = 0 THEN OUT CP,1 : GOTO WWL
        END
```

Wire CI_3 (MS) low, and enter and run the program. Use an oscilloscope to determine how long a control-out bit (say CO_0) is low which corresponds to the length of the polling loop. To be entirely safe, do not use the straight polling technique unless the hardware action is five times longer.

For a 12.5 MHz PC-AT compatible using the program above in a stand-alone compiled form, the loop time is 6 μsec.

3.5.5 Data Input/output

The final and most straightforward software operation is to input and output 8-bit values through the data ports. Devices such as ADC's, memories and programmable interval timers produce data. In the parallel approach, signals are sent from control-out which instruct the device to make a value available to the data-in port. The value is read by the program segment below.

```
        . . .
        program sends read control signals to device
        DVALUE = INP(DP)
        data value processed
        . . .
```

where DP = &H301 (for the PC ports of Section 3.1).

Data output is necessary to load values into measurement circuit memory and to program counter/timers. As in the case of inputting data, outputs occur in the context of sending appropriate control signals to the device receiving the data. A typical program segment follows the form below.

```
...
program sends write control signals to a device
OUT DP,DVALUE
...
```

3.5.6 Troubleshooting a Circuit Connected to the Parallel Ports

Consider a flip-flop wired to control-out bits 0 and 1 as shown in Figure 3.22. The signal S/S (Start/Stop) originated by the flip-flop is controlled by software as shown in Figure 3.23. CO_0 (STA*) and CO_1 (STO*) are ordinarily inactive. A brief activation of STA* makes S/S high, and a brief activation of STO* makes S/S low.

Knowledge of 7474 flip-flop operations and connections is assumed. Review Section 2.4.3 if necessary.

Software which activates STA* and STO* requires these constants.

CO_7	CO_6	CO_5	CO_4	CO_3	CO_2	CO_1	CO_0	hex	constant
X	X	X	X	X	X	STO*	STA*	value	name
1	1	1	1	1	1	1	1	FF	NORM
1	1	1	1	1	1	1	0	FE	STA
1	1	1	1	1	1	0	1	FD	STO

The following program repeatedly makes S/S alternate between high and low.

```
Program TEST S/S (dynamic)
DEFINT A-Z
CP = &H300
```

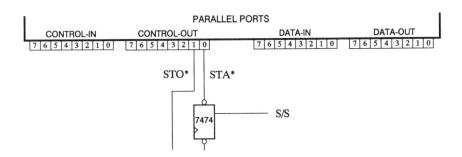

Figure 3.22. A flip-flop wired to the parallel ports and controlled by software.

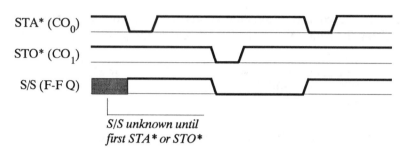

STA* (CO$_0$)

STO* (CO$_1$)

S/S (F-F Q)

*S/S unknown until
first STA* or STO**

Figure 3.23. Timing diagram illustrating software control of S/S in Figure 3.22.

```
        NORM = &HFF
        STA  = &HFE
        STO  = &HFD
REPT:  OUT CP,NORM
        OUT CP,STA          'S/S goes high
        OUT CP,NORM
        OUT CP,NORM
        OUT CP,NORM         'brief delay
        OUT CP,NORM
        OUT CP,NORM
        OUT CP,STO          'S/S goes low
        OUT CP,NORM
        GOTO REPT
        END
```

The three outputs between STA* and STO* produce a short delay which may be lengthened or shortened by adding or deleting OUT statements as desired. (A QuickBASIC program in an infinite loop may be exited by typing CONTROL-BREAK.)

This example illustrates software control of a circuit wired to the parallel ports and provides a framework to discuss troubleshooting. The flip-flop and program's operation may be verified by connecting oscilloscope channel 1 to STA* and channel 2 to S/S. STO* and S/S may also be observed.

If the circuit is not working properly, noise may be the problem. After the usual checks, wire the flip-flop's unused CLOCK input either high or low. It's possible that noise may make the flip-flop "see" CLOCK going high. If problems persist, put .01μf capacitors between STA* and ground and between STO* and ground. The capacitors act as a filter and eliminate noise spikes which might erroneously alter S/S. After the addition, STA*, STO* and S/S should be observed on the scope. It's possible that too large a capacitor might "filter" STA* or STO* to the point where intended actions are not seen by the flip-flop. Noise control capacitors are recommended in several later circuits.

Using the circuit and program above on a 12.5 MHz PC-AT compatible, the only problem occurred when another control-out bit was connected to the flip-flop's DATA input. The solution was to wire CLOCK low.

If the dynamic procedure above shows no problems, another test is to stop the program after each change in S/S and observe with a logic probe whether or not the change occurred. The modified program is.

```
'       Program TEST S/S (static)
        DEFINT A-Z
        CP = &H300
        NORM = &HFF
        STA = &HFE
        STO = &HFD
REPT:   OUT CP,NORM
        OUT CP,STA
        OUT CP,NORM
        INPUT "S/S should be high.  <Ent> "; A$
        OUT CP,NORM
        OUT CP,STO
        OUT CP,NORM
        INPUT "S/S should be low.  <Ent>  "; A$
        GOTO REPT
        END
```

If an incorrect state is observed, wire CLOCK high or low and add noise capacitors.

It's important in more complex circuits to envision static and dynamic tests for noise and other problems.

Exercises

IBM Ports

1. Design a PC port number decoding circuit which responds to 310 and 311 hex.

2. Use information on dip switches in Section 2.5.3 to design a ports circuit which responds to 0011 000 $S_3S_2$00 and 0011 0000 $S_3S_2$01 where S_3S_2 are the settings of two switches. Use EXCLUSIVE-OR gates.

3. Design a PC expansion slot circuit with four 8-bit input and four 8-bit output ports.

Apple II Ports

4. Design an Apple II expansion slot circuit with four 8-bit input and four 8-bit output ports.

Software (Most Exercises Require Parallel Ports)

5. After acquiring appropriate ports, enter and run the test programs in Sections 3.5.1 and 3.5.2.

6. Write a program which allows investigation of what happens when a number greater than 255 is output to the data port. What does happen?

7. Enter the program in Section 3.5.3 and observe its operation.

8. Measure the time between outputs as described in Section 3.5.3 for the computer and programming language being used.

9. Write a program which makes CO_7 and CO_6 undergo the timing pattern shown in Figure 3.24.

10. Devise a circuit and write a program which sets a 7474 flip-flop, waits for the signal on CI_0 (originating from a breadboard system logic switch) to go high and then resets the flip-flop.

11. Carry out the polling exercise in Section 3.5.4.

12. Use the method described in Section 3.5.4 to determine the approximate loop time for the example program.

13. Build the circuit in Section 3.5.6 Enter the first test program and observe all signals on an oscilloscope. Even if the circuit works, observe the effects of wiring CLOCK high or low and adding noise capacitors.

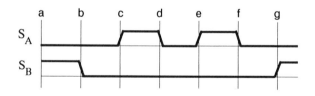

Figure 3.24. Drawing for Exercise 9.

4 Tutorial on the AD573 ADC

The approach to data acquisition described in this book utilizes three categories of large scale integrated circuits: Analog-to-Digital Converters (ADC's), Random Access Memories (RAM's), and Programmable Interval/Timers (PIT's). The specific IC's are the Analog Devices AD573 10-bit ADC, the 6264 8K by 8-bit RAM and the Intel 8253 PIT. The purpose of this and the following two chapters is to describe these integrated circuits in detail, to show how they can be interfaced to the parallel ports of Chapter 3 and to present all hardware and software for example applications. While these tutorials focus on particular devices, the ideas and principles can be applied to other ADC's, memories and PIT's, as well as to other types of devices (including voltage-to-frequency converters, digital-to-analog converters, programmable gain amplifiers, digital switches, etc.).

A variety of sensors produce voltages precisely related to such physical quantities as temperature, pressure, light intensity, current, etc. For a computer to get information from sensors, voltages must be converted to digital codes, something most efficiently done by an ADC. While this chapter only shows how to interface and control an ADC through the parallel ports, several references in Appendix A provide general discussions of sensors, operational amplifiers and the internal operation of converters.

ADC's are characterized by an input voltage range, conversion time and resolution (the number of binary bits in the converted value). The Analog Devices AD573JN is a $20, 10-bit, 20 μsec converter which accepts input voltages in one of two ranges, bipolar -5 to +5 or unipolar 0 to +10. The AD573 was chosen because its speed and 10-bit resolution are needed for applications in later chapters. However, a less expensive, slower 8-bit ADC could be substituted in this chapter.

The popular $3, 8-bit, 100 μsec ADC0808 from National Semiconductor is a good substitute. Using specifications in National's Linear Databook and the material in this chapter, it's straightforward to interface the ADC to the parallel ports and write software for the example application given in Section 4.6.

4.1 Operation and Interface

4.1.1 AD573 Integrated Circuit

Upon command, the AD573 converts an input voltage to a proportional 10-bit binary value. The following table gives the values produced by different inputs when the ADC is wired for 0 to 10 volt operation.

input voltage	10-bit converted value
0.00	00000000 00
0.01	00000000 01
0.02	00000000 10
4.99	01111111 11
5.00	10000000 00
9.98	11111111 10
9.99	11111111 11

In general, voltage = (10-bit value/1024) * 10 volts. Resolution, the smallest voltage difference which can be measured, is 10 mv. Accuracy for the ADC573JN is ±LSB (least significant bit), meaning converted values are determined to (±1/1024)*10 = ±.01 volts.

It's possible to read just the upper 8 bits (high byte) of a converted value. In this case, voltage = (8-bit value/256) * 10 and resolution is .04 volts.

Connections to the 20 pin IC, shown in the logic drawing in Figure 4.1, are:

connection	pin	function
V	14	Input for the Voltage to be converted.
AG	15	Analog (voltage) Ground
V-	13	-12 to -15 Volt power supply
V+	11	+5 Volt power supply
DG	17	Digital Ground for the +5 supply, control signals and data lines
BO	16	Bipolar Offset (ground for unipolar 0 to 10 range, open for bipolar -5 to +5)

Figure 4.1. Logic drawing of the Analog Devices' AD573. A pin assignment drawing is in Appendix C.

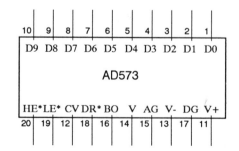

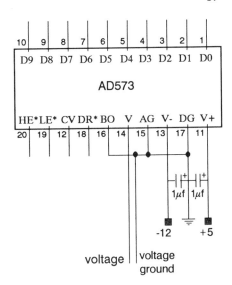

Figure 4.2. Power supply, ground, range and input voltage connections to an AD573.

CV	12	ConVert input initiates conversion
DR*	18	Data Ready output from the ADC indicates when conversion starts and stops
HE*	20	High byte Enable activation causes the ADC to put the high 8 bits of a converted value on data lines D9...D2
LE*	19	Low byte Enable activation causes the ADC to put the 2 low bits on D1 and D0
D9...D0	10...1	Data lines are in a disconnected Hi-Z state unless HE* or LE* is active

4.1.2 Analog Setup

Figure 4.2 shows the AD573 setup for 0 to 10 volt operation (BO is grounded). The input voltage and its ground go to V and AG. +5 and -12 volt power supplies are connected as shown. Although not required, the analog and digital grounds are connected and 1 μf tantalum capacitors are put between the power supply voltages and ground.

Analog Devices' *Data-Acquisition Databook* presents methods for precise calibration, bipolar operation and connecting a sample and hold, as well as information on different AD573 grades.

4.1.3 Control and Timing

In this book, the three major integrated circuits are individually interfaced to the parallel ports and controlled by software. To design the interface and write the software, it's first necessary to understand the IC's operation. Figure 4.3 gives timing specifications for the AD573.

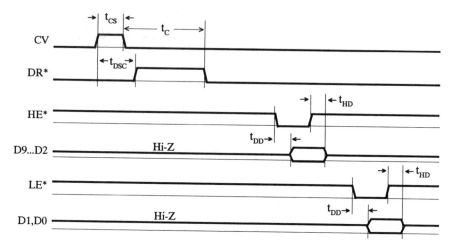

Figure 4.3. Timing and control specifications for the AD573.

Conversion is initiated when the ADC receives a positive CV pulse. The pulse must remain active a minimum of $t_{CS} = 500$ ns. The chip responds to CV's leading edge by deactivating DR* (data from the previous conversion is available until DR* goes inactive). This may take as long as $t_{DSC} = 1.5$ μsec. The chip responds to the trailing edge of CV by beginning conversion. When conversion is complete, the chip takes DR* active. Conversion time t_C varies from 10 to 30 μsec.

After conversion, the ADC responds to HE* activation by putting the high 8 bits of the converted value on D9...D2. Data is stable t_{DD} seconds after activation and remains t_{HD} seconds after HE* deactivation. Similarly, LE*'s activation causes the ADC to put the low 2 bits of the converted value on D1 and D0. Maximum t_{DD} is 250 ns and minimum t_{HD} is 50 ns.

4.1.4 Parallel Ports Interface

With Figure 4.3 in mind, how can an AD573 be interfaced to the parallel ports so software can produce CV, monitor DR* and use HE* and LE* to read data?

Suppose the following bits of the control-out port are connected to the ADC:

CO_3 to   CV
CO_4 to   HE*
CO_5 to   LE*

With these connections, conversion is initiated by software outputting a sequence of bytes to control-out which make CV go low-high-low.

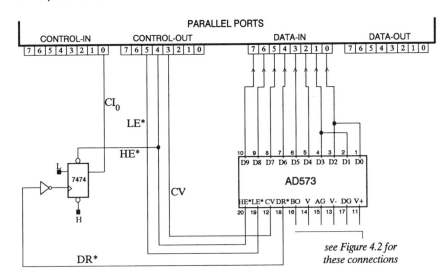

Figure 4.4. Complete circuit for an AD573 interfaced to the parallel ports.

Software then must wait until data is ready. One way to do this is to connect DR* to a control-in bit and have software loop as long as the bit is high. However, the approach has a flaw. The time between software producing CV and beginning the wait loop depends on the speed of the computer and language. If the time is extremely fast, DR* might not yet be high as shown in Figure 4.3. On the other hand, if the time is longer than 20-30 μsec, DR* might already be back low after conversion. Therefore, waiting as long as high would produce in the first case a premature end of the loop and in the second an infinite loop. So how can software reliably wait until data is ready? The answer is more hardware.

Consider the circuit in Figure 4.4. The output of a 7474 flip-flop is connected to CI_0. HE* is connected to the 7474's PRESET* input. The flip-flop's DATA input is wired low and the inverse of DR* goes to the positive edge triggered CLOCK input. Only activation of HE* makes the flip-flop high, and only DR* going low (CLOCK high) makes the flip-flop low. With this circuit, software briefly activates HE* (makes CI_0 high), sends CV, waits while CI_0 is high, and finally inputs the converted value (when DR* makes CI_0 low). Activation of HE* to input data also puts CI_0 back high and sets the stage for the next conversion.

Also in Figure 4.4, ADC data lines D9...D2 are connected to the 8 bits of the data-in port. To read the high byte of the converted value, software activates HE*, inputs the high byte from the data port, and deactivates HE*. If desired, the low 2 bits are read by activating LE*, inputting from data-in

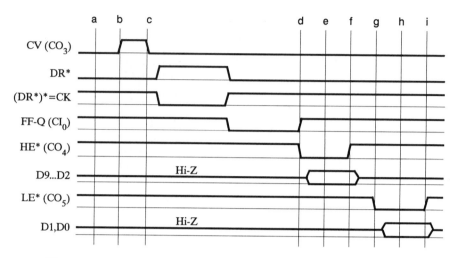

Figure 4.5. A complete timing diagram for an AD573 interfaced to the parallel ports. The letters at the top label software actions.

and deactivating LE*. With the ADC's D1 and D0 connected to DI_1 and DI_0, only the low 2 bits of the input value matter.

Figure 4.5 is a complete timing diagram showing the relationships among the ADC, flip-flop and software.

4.2 Software

An intimate relationship exists between connecting the ADC to the parallel ports and envisioning software which operates the chip. With the interface complete, the software is presented.

A reasonable approach is to divide software into the five operations summarized below.

4.2.1 Initialize the Flip-flop

The flip-flop must be set before each conversion. Since HE* is connected to PRESET*, reading data from one conversion accomplishes this for the next. But what about the first? The following sequence of outputs, sent to the control port before the first conversion, briefly activates HE* and sets the flip-flop.

CO_7	CO_6	CO_5	CO_4	CO_3	CO_2	CO_1	CO_0	hex	constant
X	X	LE*	HE*	CV	X	X	X	value	name
1	1	1	1	0	1	1	1	F7	NORM
1	1	1	0	0	1	1	1	E7	READH
1	1	1	1	0	1	1	1	F7	NORM

The table above, as usual, shows the relationship between control-out bits and the circuit. X denotes unconnected bits. The practice in this book is to output 1 for bits that don't matter. The constants and hex values are for later use in actual routines. The value of the NORMal, inactive state is F7 since ConVert is ordinarily low.

4.2.2 Initiate Conversion

The following outputs send a convert pulse to the ADC while keeping the other controls inactive. Points in the timing pattern of Figure 4.5 are also given in the table.

point	CO_7 X	CO_6 X	CO_5 LE*	CO_4 HE*	CO_3 CV	CO_2 X	CO_1 X	CO_0 X	hex value	constant name
a	1	1	1	1	0	1	1	1	F7	NORM
b	1	1	1	1	1	1	1	1	FF	CONVERT
c	1	1	1	1	0	1	1	1	F7	NORM

Time between outputs might be sufficiently short that CV is not high the required 500 ns. In Section 3.5.3, the time was determined (for the particular computer, language and ports being used). If necessary, output CONVERT a second or third time.

4.2.3 Wait While CI_0 Is High

After sending convert, software waits while CI_0 is high. The following table shows how to determine whether CI_0 is 0 or 1. CP is the control port number. X means a bit is arbitrarily high or low.

	CI_7 X	CI_6 X	CI_5 X	CI_4 X	CI_3 X	CI_2 X	CI_1 X	CI_0 FF-Q
INP(CP)	X	X	X	X	X	X	X	0 or 1
MASK	0	0	0	0	0	0	0	1
INP(CP) AND MASK	0	0	0	0	0	0	0	0 or 1

If the result of the AND operation equals MASK, CI_0 is high; and if the result eqauls zero, CI_0 is low.

4.2.4 Input the High Data Byte

When data is ready, the following sequence inputs the high byte. Again, the points refer to Figure 4.5.

point	CO_7 X	CO_6 X	CO_5 LE*	CO_4 HE*	CO_3 CV	CO_2 X	CO_1 X	CO_0 X	hex value	constant name
c	1	1	1	1	0	1	1	1	F7	NORM
d	1	1	1	0	0	1	1	1	E7	READH

point									hex value	constant name
e	input from the data port (value = D9...D2)									
f	1	1	1	1	0	1	1	1	F7	NORM

The ADC does not put data on D9...D2 until 250 ns after HE* goes active. For very fast computers, the input from the data port may occur before 250 ns. Again, the problem is solved by knowing I/O times from Section 3.5.3, and, if necessary, adding a second or third READH step.

4.2.5 Input the Low Data Bits

After reading the high byte, the low 2 bits are input as follows.

point	CO_7 X	CO_6 X	CO_5 LE*	CO_4 HE*	CO_3 CV	CO_2 X	CO_1 X	CO_0 X	hex value	constant name
f	1	1	1	1	0	1	1	1	F7	NORM
g	1	1	0	1	0	1	1	1	D7	READL
h	input from the data port (value = 111111,D1,D0)									
i	1	1	1	1	0	1	1	1	F7	NORM

Since D9...D4, connected to $DI_7...DI_2$, are in a Hi-Z state, the high 6 bits in the input value are 1 (because in TTL logic a disconnected input is "seen" as high).

4.2.6 Constants

Constants in the five operations above are:

NORM = &HF7	**NORM**al, inactive state of control outputs
CONVERT = &HFF	**CONVERT** pulse to ADC's CV input
MASK = &H01	**MASK** for control-in which only passes CI_0
READH = &HE7	**READ H**igh data byte from ADC by activating HE*
READL = &HD7	**READ L**ow data bits by activating LE*
CP	Control Port number (for an IBM PC with the parallel ports in Section 3.2, CP=&H300)
DP	Data Port number (for the PC ports in Section 3.2, DP=&H301)

4.2.7 Subroutines and Wait Loop

The following four subroutines and wait statement use the constants above to carry out the operations in Sections 4.2.1-4.2.5.

```
'        Subroutine to INITialize the flip-flop
         SUB INIT
            OUT CP,NORM
```

```
        OUT CP,READH
        OUT CP,NORM
     END SUB
```

'   Subroutine to **START** Conversion

```
     SUB STARTC
        OUT CP,NORM
        OUT CP,CONVERT
        OUT CP,CONVERT
        OUT CP,NORM
     END SUB
```

'   Subroutine to **READ** the **HIGH** data byte

```
     SUB READHIGH(HIGHB)
        OUT CP,NORM
        OUT CP,READH
        OUT CP,READH
        HIGHB = INP(DP)
        OUT CP,NORM
     END SUB
```

'   Subroutine to **READ** the **LOW** data bits

```
     SUB READLOW(LOWB)
        OUT CP,NORM
        OUT CP,READL
        OUT CP,READL
        LOWB = INP(DP)
        LOWB = LOWB AND &H03
        OUT CP,NORM
     END SUB
```

'   Statement which Waits While CI0 is High

```
WWH:  IF (INP(CP) AND MASK) = MASK THEN GOTO WWH
```

The result of the AND operation equals MASK if CI_0 is high when the input occurs.

In order to use these routines and the wait loop, the constants must be given values. The approach adopted for software in this book is that all main programs make constants integer, assign them values and declare them shared with subroutines. The following QuickBASIC statements, at the beginning of main programs, implement the approach for AD573 software.

```
     DEFINT A-Z
     COMMON SHARED CP,DP
     COMMON SHARED NORM,CONVERT,READH,READL,MASK
     CP = &H300 : DP = &H301
     NORM = &HF7
     CONVERT = &HFF
     READH = &HE7 : READL = &HD7
     MASK = &H01
```

4.3 Example Application - A Voltmeter

A digital voltmeter is easily implemented with the ADC interfaced as shown in Figure 4.4. The following program repeatedly obtains and displays 10-bit voltage values.

```
'       Program VOLTMETER
        DEFINT A-Z
        COMMON SHARED CP,DP
        COMMON SHARED NORM,CONVERT,READH,READL,MASK
        CP = &H300 : DP = &H301
        NORM = &HF7
        CONVERT = &HFF
        READH = &HE7 : READL = &HD7
        MASK = &H01
        CLS
        PRINT "Voltage is Continually Read and Displayed"
        PRINT "Type Any Key to Abort"
        CALL INIT
MAINL: CALL STARTC
WWH:   IF (INP(CP) AND MASK) = MASK THEN GOTO WWH
        CALL READHIGH(HIGHB)
        CALL READLOW(LOWB)
        VALUE = HIGHB*4 + LOWB
        VOLT! = (VALUE/1024) * 10
        LOCATE 10,10
        PRINT "Volts = ";
        PRINT USING "##.##";VOLT!
        IF INKEY$ = "" THEN GOTO MAINL
        END

        Text of INIT, STARTC, READHIGH and READLOW
        subroutines.
```

In the program, the 10-bit converted value is constructed after the high and low bytes are input. Since D9...D2 starts in the 2^2 place,

```
        VALUE = HIGHB*4 + LOWB
```

To see that this is correct, suppose D9...D0 is 11111111 11 = 1023. HIGHB = 11111111 = 255 and LOWB = 00000011 = 3. So, VALUE = 255*4 + 3 = 1023.

The actual voltage is computed as discussed in Section 4.1.

```
        VOLT! = (VALUE/1024) * 10
```

The appended "!" overrides DEFINT A-Z and makes VOLT a single precision real variable.

4.4 Troubleshooting Voltmeter Hardware and Software

Construct on a breadboard the voltmeter circuit in Figure 4.4. Refer to the pin assignments in Appendix C and the analog setup in Figure 4.2 to carefully wire the ADC's bipolar offset input to the analog and digital grounds and to the power supply ground(s). Connect -12 and +5 volts as shown. Verify the voltages with a traditional voltmeter. Check all +5 and ground connections with a logic probe. Put the probe directly on IC pins to eliminate the possibility of bent pins or a broken wire. Enter, save and print the complete voltmeter program. Proof and correct the text. Apply a known DC voltage (between 0 and 10 volts) to ADC pin 14. Run the program. Despite all precautions, don't expect the system to work. It's likely that something is wrong with the software, parallel ports or hardware. The following is a reasonable troubleshooting sequence.

4.4.1 Check the Parallel Ports and Software

Individually verify each of the four subroutines and the monitor procedure as follows. Load the voltmeter program. Eliminate program lines from CLS through IF INKEY$. Be sure to keep the constants and subroutines. Then systematically and thoroughly carry out the following tests.

Test INIT

Replace the eliminated program lines with this loop:

```
REPT:  CALL INIT
       GOTO REPT
```

With the program running, use an oscilloscope to see if HE* (CO_4) is repeatedly activated. A good practice is to put the scope probe directly on the flip-flop's PRESET* pin and the ADC's pin 20.

If HE* is not cycling, disconnect the signal from the ADC and flip-flop and check again. If HE* is now as expected, reconnect the flip-flop and then the ADC to determine which chip is crashing the control. Replace the offending IC and recheck the wiring.

If HE* is not repeatedly activating after disconnection, either the software or parallel ports are not working. Test the ports with programs from Chapter 3. At this point, it's essential to known the ports including cables are functioning properly.

If the ports check out, the problem is the INIT subroutine and associated constants. The most common errors are misspelled constant and subroutine names as well as incorrect constant values.

If no errors are found and HE* is still not activating, save test INIT and enter the following new program.

```
'        Program EMERGENCY
REPT:  OUT &H300,&HF7              'output NORM
       OUT &H300,&HE7              'make HE* (CO4)low
       OUT &H300,&HF7              'return HE* high
       GOTO REPT
       END
```

Run the program. If HE* cycles as expected, restore test INIT and repeat the tests above. If the emergency program does not produce a correct HE*, carefully check the program and recheck the parallel ports.

Even if test INIT isolates a problem which after correction results in a working HE*, it's still a good idea to continue the troubleshooting procedure before going back to the complete voltmeter system.

Recall that an infinite QuickBASIC loop is exited by typing CONTROL-BREAK.

Test STARTC

Next, test the START Convert subroutine by replacing the CALL INIT loop with

```
REPT:  CALL STARTC
       GOTO REPT
```

Check with an oscilloscope that the signal ConVert (ADC pin 12) is ordinarily low but briefly goes high. If not, disconnect it from the ADC and check again. If necessary, check the parallel ports and then the software (as described above for INIT).

If CV is correctly cycling, check the ADC's Data Ready output. Put CV on scope channel 1 and DR* (ADC pin 18) on channel 2. Does DR* follow the pattern in Figure 4.3? If not, be sure there is enough time (30 μsec) between CV's. Extend the loop by adding statements between CALL and GOTO. Next, check that the length of CV meets the minimum 500 ns requirement. If not, add extra "out convert" statements to the subroutine and check again. If CV is long enough but there is still no DR*, replace the ADC chip and carefully check the power supply voltages. If necessary, rewire the circuit on a new breadboard.

Test READHIGH and READLOW

Next, test the subroutines READHIGH and READLOW by calling them in a loop and observing HE* and LE* with an oscilloscope. If something is wrong, follow the procedure above for INIT and STARTC.

Also, after activating HE* or LE*, there must be at least 250 ns before data is read. Add extra out statements to the subroutines if this may be a problem.

Test the Monitor Loop

The final software test is wait while CI_0 is high. Disconnect the flip-flop and wire CI_0 to an active-low breadboard system pulser. Load the original voltmeter program and change the IF statement as follows:

```
WWH:   IF (INP(CP) AND MASK) = MASK THEN
           INPUT "CI0 now low "; A$
           GOTO WWH
       ENDIF
```

Start the program. Activate the pulser and see if the input message occurs. If not, check the parallel ports and software as outlined above. If the message occurs, while program execution is stopped awaiting input, check that the flip-flop output is low.

> Recall that a QuickBASIC program can be aborted by responding to an input with CONTROL-C.

If the parallel ports and software meet the tests above yet the voltmeter still doesn't work, then the problem is isolated to the hardware.

4.4.2 Check the ADC and Flip-flop

Modify the original voltmeter program by removing the lines between VALUE = and PRINT USING. This speeds up the data collection loop for oscilloscope observation. Start the program with scope channel 1 looking at CV (CO_3). Check the signals in the timing pattern in Figure 4.5: 1) DR*; 2) (DR*)* equals the flip-flop CK input; 3) the flip-flop output (CI_0); 4) HE*; and 5) LE*. If any signal is wrong, carefully check everything it is connected to.

If CV and DR* are right and CI_0 is wrong, the problem may be that noise inadvertently sets or clears the flip-flop. Put a .01 μf noise capacitor between PRESET* and ground. Afterward, be sure HE* still meets the length requirements. Although less likely to be a problem, look for noise on the flip-flop's CLOCK and DATA inputs.

If, on the other hand, the five signals follow CV as expected, the last thing to check is the ADC's data-out bits. Trigger the scope on HE*'s high-low transition and successively test each output bit on channel 2. Observe the transition from Hi-Z to either a normal high or low. If the voltage applied to the ADC is constant, the high 8 bits should be the same each scope sweep.

Also, test the low 2 bits by triggering on LE*'s activation. The bits probably will not have the same value each scope sweep, unless noise and other variations are less than 10 mv.

The software might not be fast enough to get "good" scope patterns for these tests. Another approach is to stop program execution with HE* or LE* active. This is accomplished by putting an input statement in the READHIGH and READLOW subroutines. Then, when program execution stops with HE* or LE* active, a logic probe could be used to check each ADC data bit.

If the ADC is outputting reasonable values but the voltage is wrong when displayed by software, then it's likely that 1 or more data bits are wired to the wrong lines of the data-in port.

4.4.3 Summary

In systems where many things could be wrong, the objective of troubleshooting is to establish subsets of the overall operation which are definitely working.

4.5 Conclusions

The simple voltmeter program tests the ADC, interface circuit and software. However, additional capabilities are necessary for reasonable systems.

It's desirable to be able to convert voltages at a programmable rate (up to some maximum). The counter/timer presented in Chapter 6 makes this possible. A complete programmable rate voltage measuring system is given in Chapter 9.

As long as software inputs values after each conversion, the maximum rate of acquisition is insufficient for many applications. Fast voltage measurement is achieved by hardware converting and storing values in the memory IC presented in the next chapter. A complete fast voltage measurer is given is Chapter 10.

This chapter illustrates the interplay among a device, a support circuit, the parallel ports, and control and monitor software. While more complex devices with more complex circuits and software follow, the basic approach remains the same. First, acquire a clear picture of the device and how it works. Then design hardware and envision software which allow the device to be interfaced and operated.

Exercises

1. Use parallel ports (which meet the specifications in Chapter 3) and a breadboard system to wire an AD573 to the parallel ports as shown in Figure 4.4. IC pin assignments are in Appendix C. Enter the software for the voltmeter and test the system by applying known voltages. Even

if the system works, go through the troubleshooting sequence in Section 4.7.

2. Devise a circuit and software to input voltages from two AD573's.

3. Write a voltmeter program which only inputs the high 8 bits of the converted value.

4. What is the accuracy of 8-bit voltage values if conversion is to ±LSB (of the 10-bit converted value)?

5. Write a program which obtains and stores 8-bit voltage values every .10 seconds a selected number of times. Use software loops, a stopwatch, and trial and error.

6. An interface circuit for the ADC0808 is in Figure 4.6. Use National Semiconductor's *Linear Databook* to implement a digital voltmeter.

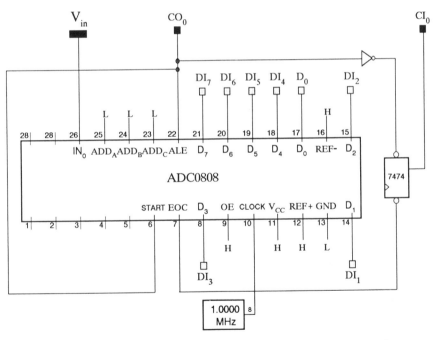

Figure 4.6. Interface for ADC0808 and the parallel ports for use in Exercise 6.

5 Tutorial on the 6264 RAM

Memory is the second building block for measurement circuitry. In the parallel approach described so far, the maximum rate of data acquisition is determined by how fast the host computer can poll for data ready, input 1 or 2 bytes, carry out housekeeping operations (such as determining when to terminate acquisition), and then resume polling for the next data. For the ADC circuit in Chapter 4, with the $20\,\mu$sec ADC and a 12.5 MHz PC AT compatible using Microsoft QuickBASIC, the maximum rate of converting voltage and inputting and storing 8-bit values is around 2000 readings/second. The capability for faster rates is desirable. A way to accomplish this is for the hardware to temporarily store data in its own memory. Then, after acquisition, the values are read by the host computer. This chapter introduces a particular memory chip, shows how to interface it to the parallel ports, presents control software, and gives a detailed application. Later, Chapter 10 describes a fast voltage measurer which uses memory.

A memory integrated circuit contains a set of storage locations each with a unique address. To write a value to a location, the address is specified, the value is made available, and the chip is told to store the value. To read a previously stored value, the address is specified and the chip is instructed to make the value available. Memory IC's are characterized by the number of storage locations, the bit size of stored values, and the control protocol.

The 8K by 8-bit 6264 static random access memory has been selected for temporary data storage. It is produced by several manufacturers and comes in a variety of models and speeds.

5.1 Operation

The 6264 is a 28 pin integrated circuit with 8192 addressable locations at which 8-bit values may be randomly stored and accessed. The IC is shown in Figure 5.1.

The connections are:

$A_{12}...A_0$ 13 Address lines (which allow addresses from binary 0 0000 0000 0000 to 1 1111 1111 1111 or from hex 0000 to 1FFF or from decimal 0 to 8191).

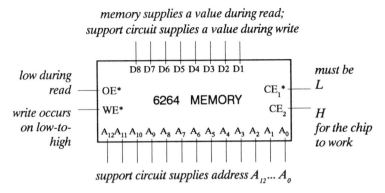

Figure 5.1. Logic drawing of the 6264 8K by 8-bit static RAM. Pin assignments, including +5 and ground, are in Appendix C.

D8...D1   8 bi-directional **Data** lines.

CE_1* and CE_2 **Chip Enable** controls must be active for the chip to respond to the other commands.

WE*   **Write Enable** (the chip stores the value on its data lines when WE* returns inactive).

OE*   **Output Enable** (when active, the chip puts the value at the current address on the data lines; when inactive, the data outputs are in a Hi-Z state).

To use the 6264, a support circuit must generate 13 address lines, a bi-directional 8-bit data bus and the four control lines. Circuit complexity depends on the details of the application. For temporary data storage, it's only necessary to write to successive addresses and afterward read from successive addresses. Specifically:

	binary	address hex	decimal
first data value at	0 0000 0000 0000	0000	0
next value at	0 0000 0000 0001	0001	1
next at	0 0000 0000 0010	0002	2
...	...	...	...
next to last possible value at	1 1111 1111 1110	1FFE	8190
last possible value at	1 1111 1111 1111	1FFF	8191

Notice that the succession of addresses is identical to the outputs of binary counters.

The chip enables are used when multiple 6264's are present. In this chapter, only one chip is needed and CE_1* and CE_2 are wired active. In Chapter 10, four 6264's are managed by the CE's.

In order to interface a 6264 to the parallel ports, the chip's read and write timing sequences must be understood.

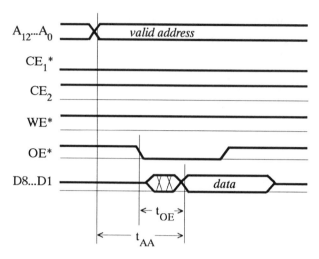

Figure 5.2. Timing sequence for reading the value stored at a specific 6264 address.

5.1.1 Read from an Address

Figure 5.2 shows how to read the value stored at a specific address. The chip enables are active and WE* is inactive. The 6264 makes available the value stored at an address t_{AA} seconds after the address is supplied to $A_{12}...A_0$. However, it's not until t_{OE} seconds after OE* goes active that the value is actually put on the data lines.

6264 IC's are specified by speed. In the 150 ns version, maximum t_{AA} is 150 ns and maximum t_{OE} is 70 ns. Also, this speed 6264 cannot respond until 120 ns after the enables are activated. For applications in this book, none of these times are critical. However, it's always wise to consult the manufacturer's specifications for the particular speed and type of 6264 being used. References are in Appendix A.

5.1.2 Write to an Address

Figure 5.3 shows how to store a value at a specific addresses. Again, CE_1* and CE_2 enable the 6264. OE* is inactive. The chip stores the value on the data lines at the current address when WE* goes from low-to-high.

Important times, shown in the figure, are:

(1) WE* must be low at least t_{WP} seconds.
(2) The address must be stable at least t_{AW} seconds before WE* goes from low to high.
(3) The data must be stable at least t_{DW} seconds before WE* goes high.
(4) The address must remain at least t_{WR} seconds after WE* goes high.

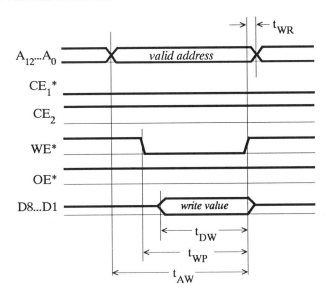

Figure 5.3. Timing sequence for writing a value to a specific 6264 address.

Minimum values of these times for a 150 ns 6264 are: t_{WP} = 90 ns, t_{AW} = 100 ns, t_{DW} = 60 ns and t_{WR} = 5 ns.

Each of these times can be much longer. Further, unlike the address, the data value can be removed from the data lines at the same time WE* goes high. For further details on write timing, consult manufacturers' specifications.

Before showing how to interface the chip to the parallel ports, address generating circuitry is explained.

5.2 Address Generator

The 6264 requires 13 address lines ($A_{12}...A_0$). With read and write sequences limited to successive addresses, only two operations are required: zero the address and increment the address by one.

The outputs of four 74193 binary counters, wired as shown in Figure 5.4, produce the 13 address lines.

Recall that the 74193 count (Q_D, Q_C, Q_B and Q_A) increases by one when its CU (Count Up) input goes from low to high. The CA (CArry) output goes from low to high when the count rolls over from 1111 to 0000, and, if connected to the next higher counter's CU, causes an increment. The CL (CLear) to zero control must be low for a 74193 to count. When the LD* (Load Data)

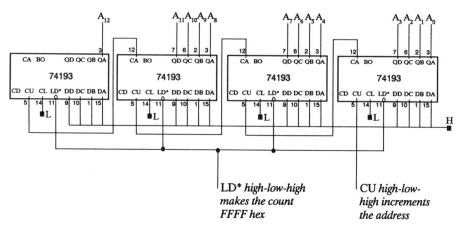

Figure 5.4. Thirteen-bit address generator using 74193 binary counters.

input is activated, counting stops and the count value equals the Data inputs (D_D, D_C, D_B and D_A).

In Figure 5.4, the signal LD* (LoaD address counters) is connected to the four counters' LD*'s. All data inputs are wired high. The signal CU (address Count Up) is connected to the CU of the low ($A_3...A_0$) counter. CA from this counter goes to CU of the next higher counter, and so on. The two required operations are:

To Zero the Address. Make the address lines all high by briefly activating LD*. Then, make the address lines all zero by taking CU high-low-high causing the low counter to roll over to 0000, producing a carry which causes the next higher counter to roll over, and so on.

To Advance the Address. Advance the address by taking CU high-low-high, incrementing the low counter and producing carries as needed.

The reason for CU to go high-low-high as opposed to the equally workable low-high-low is the desirability for the inactive state of all controls to be high. Also, the 74193 CL's could be used to zero the counters in one step. However, considerations in the design of an example later in this chapter and in the fast voltage system in Chapter 10 dictate the LD, CU sequence to zero the address.*

5.3 Parallel Ports Interface

Before giving an example of fast data acquisition, the memory and address generator are interfaced to the parallel ports so software can write to and read from successive addresses. To accomplish this, it's necessary to set up a bi-directional data bus.

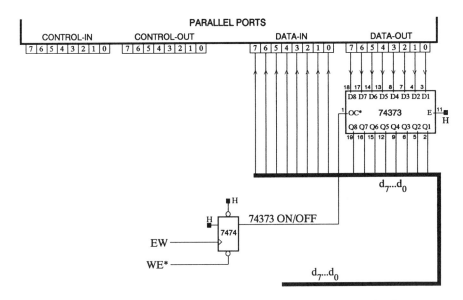

Figure 5.5. Circuit to produce a bi-directional data bus $d_7...d_0$. Data-out is put on the bus when WR* goes active and taken off when EW goes high.

5.3.1 Bi-directional Data Bus

The memory IC has bi-directional data connections D8...D1. During write operations, values go from the data-out port to D8...D1. During reads, values go from D8...D1 to data-in. Therefore, it's necessary to set up the bi-directional data bus shown in Figure 5.5.

Two signals control bus operations. Activation of the memory's WR* clears the flip-flop and turns the 74373's OC* on. The condition remains until EW (End Write) goes low-high, thereby setting the flip-flop and turning OC* off.

For the memory write operation shown in Figure 5.3, WR* could be connected directly to OC*. However, other memory IC's as well as other category devices which have bi-directional data connections require data to be available for a time after the write control returns inactive. Hence the need for a separate off.

The data bus is continually connected to data-in. So whenever the memory puts a value on $d_7...d_0$, it's available to the port. Of course, this must never occur when OC* is active.

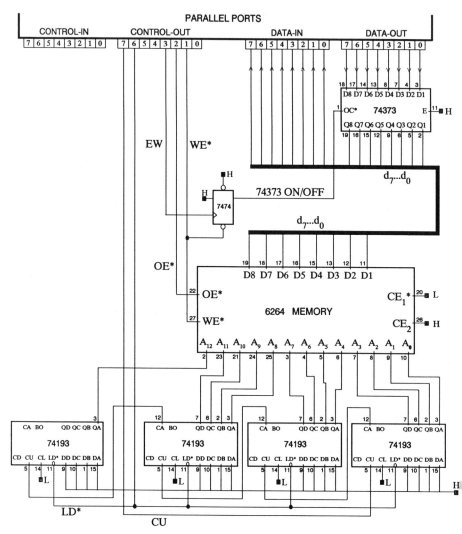

Figure 5.6. Circuit showing a 6264 memory and address generator interfaced to the parallel ports.

5.3.2 Interface Circuit

Figure 5.6 gives the complete interface circuit which enables software to operate the memory through the following 5 control-out bits.

CO_1 WE*   memory Write Enable supplies the 6264's WE* and, by clearing the 7474 flip-flop, turns the 74373 on by activating its OC*

CO_2	OE*	memory Output Enable supplies the 6264's OE*
CO_3	EW	End memory Write turns the 74373 off by deactivating its OC*
CO_6	LD*	LoaD address counters
CO_7	CU	Count Up address

In order to visualize the sequences of outputs to the control-out port which operate the circuit, the table below shows the relationship between the control-out bits and the control signals. The table also establishes the normal, inactive state for the circuit.

CO_7	CO_6	CO_5	CO_4	CO_3	CO_2	CO_1	CO_0	hex value	constant name
CU	LD*	X	X	EW	OE*	WE*	X		
1	1	1	1	1	1	1	1	FF	NORM

5.3.3 Sequence to Write to Successive Addresses

The following steps utilize the address circuit and the timing sequence in Figure 5.3 to write values to successive memory addresses.

(1) Zero the address in two steps. First, set the address to all 1's by outputting a sequence of bytes to the control port which make LD* go from its normal high to low and back high while all other control bits remain in their normal state. Then roll the address over to all 0's by outputting bytes which make CU go high-low-high. With constants defined for later use in software, the sequence is:

CO_7	CO_6	CO_5	CO_4	CO_3	CO_2	CO_1	CO_0	hex value	constant name
CU	LD*	X	X	EW	OE*	WE*	X		
1	1	1	1	1	1	1	1	FF	NORM
1	0	1	1	1	1	1	1	BF	MALOAD
1	1	1	1	1	1	1	1	FF	NORM
0	1	1	1	1	1	1	1	7F	MAADV
1	1	1	1	1	1	1	1	FF	NORM

(2) Output the value to be stored to the data port.
(3) Activate WE* (with the other controls unchanged). The action clears the flip-flop and turns the 74373 buffer on.

CO_7	CO_6	CO_5	CO_4	CO_3	CO_2	CO_1	CO_0	hex value	constant name
CU	LD*	X	X	EW	OE*	WE*	X		
1	1	1	1	1	1	1	1	FF	NORM
1	1	1	1	1	1	0	1	FD	MWRITE

(4) Store the value on the data lines at the current memory address by making WE* return high. The action leaves the flip-flop output low and the 74373 still on for the next write.

CO_7	CO_6	CO_5	CO_4	CO_3	CO_2	CO_1	CO_0	hex	constant
CU	LD*	X	X	EW	OE*	WE*	X	value	name
1	1	1	1	1	1	1	1	FF	NORM

Recall from Figure 5.3 that WE* must be low t_{WP} = 90 ns (typical value). Further, the value to be written must be present on data-out t_{DW} = 60 ns before WE* returns high. For "fast" computers (as analyzed in Section 3.5.3), it might be necessary to output a second or third MWRITE to extend WE*.

(5) Advance to the next address by making CU go from its normal high to low and then back high.

CO_7	CO_6	CO_5	CO_4	CO_3	CO_2	CO_1	CO_0	hex	constant
CU	LD*	X	X	EW	OE*	WE*	X	value	name
1	1	1	1	1	1	1	1	FF	NORM
0	1	1	1	1	1	1	1	7F	MAADV
1	1	1	1	1	1	1	1	FF	NORM

As shown in Figure 5.3, the address must be stable t_{AW} = 100 ns (typically) before WE* returns inactive.

(6) Repeat steps 2 to 5 until the desired number of values are stored.
(7) After the last value is stored, remove the 74373 from $d_7...d_0$ by making EW go from its normal high to low and back high. The action uses the flip-flop's clock/data mode to return its output and thus OC* high.

CO_7	CO_6	CO_5	CO_4	CO_3	CO_2	CO_1	CO_0	hex	constant
CU	LD*	X	X	EW	OE*	WE*	X	value	name
1	1	1	1	1	1	1	1	FF	NORM
1	1	1	1	0	1	1	1	F7	MENDW
1	1	1	1	1	1	1	1	FF	NORM

5.3.4 Sequence to Read from Successive Addresses

The following steps utilize the control sequence in Figure 5.2 to read values previously stored in memory.

(1) Zero the address (step 1 in the write sequence).
(2) Turn the memory on by activating OE*. The 6264 puts the value at the current address on $d_7...d_0$.

CO_7	CO_6	CO_5	CO_4	CO_3	CO_2	CO_1	CO_0	hex	constant
CU	LD*	X	X	EW	OE*	WE*	X	value	name
1	1	1	1	1	1	1	1	FF	NORM
1	1	1	1	1	0	1	1	FB	MREAD

(3) Input from the data port and save the value supplied by the memory. As shown in Figure 5.2, value cannot be input until at least t_{OE} = 70 ns (typically) after OE* goes active. In extremely fast computers (see Section 3.5.3), a second or third MREAD output might be needed.

(4) Remove the value from $d_7...d_0$ by returning OE* high.

CO_7	CO_6	CO_5	CO_4	CO_3	CO_2	CO_1	CO_0	hex value	constant name
CU	LD*	X	X	EW	OE*	WE*	X		
1	1	1	1	1	1	1	1	FF	NORM

(5) Advance the address by making CU go high-low-high (step 5 of the write procedure).
As shown in Figure 5.2, data is not available for input until at least t_{AA} = 150 ns after the address value is stable.

(6) Repeat steps 2-5 until the desired number of values are read.

5.3.5 Constants

The constants defined in the write and read sequences are listed here.

NORM = &HFF	NORMal, inactive state
MALOAD = &HBF	Memory Address LOAD
MAADV = &H7F	Memory Address ADVance by 1
MWRITE = &HFD	Memory WRITE at current address
MENDW = &HF7	Memory END Write
MREAD = &HFB	Memory READ from current address
CP	Control Port number (for the IBM ports described in Section 3.2, CP = 300 hex)
DP	Data Port number (for IBM's, DP = 301 hex)

5.4 Software

The operations in Sections 5.3.1 and 5.3.2 are divided into the following five subroutines designed in anticipation of various main programs which use the 6264. (As always with software, other schemes may prove more useful with experience.)

ZEROADR	ZERO memory ADdRess (make $A_{12}...A_0$ = 0)
ADVADR	ADVance memory ADdRess by 1
READMEM(value)	READ MEMory and return value at current address
WRITEMEM(value)	WRITE value to MEMory at current address
ENDMEMWR	END MEMory WRite (turn the 74373 off)

The software style adopted for this book is that all main programs declare constants as integers, assign them values and make them shared with subroutines. For memory routines, this is accomplished by including the lines below at the beginning of main programs. Recall that the integer declaration, DEFINT A-Z, can be overruled by appending a special character to a constant or variable name (e.g. "!" for single precision real values).

```
DEFINT A-Z
COMMON SHARED CP, DP
COMMON SHARED NORM, MAADV, MALOAD
COMMON SHARED MWRITE, MENDW, MREAD
CP = &H300
DP = &H301
NORM = &HFF
MAADV = &H7F : MALOAD = &HBF
MREAD = &HFB
MWRITE = &HFD : MENDW = &HF7
```

5.4.1 The Subroutines

```
SUB ZEROADR            'Step 1 in Sec. 5.3.3
   OUT CP,NORM
   OUT CP,MALOAD
   OUT CP,NORM
   OUT CP,MAADV
   OUT CP,NORM
END SUB

SUB ADVADR             'Step 5 in Sec. 5.3.3
   OUT CP,NORM
   OUT CP,MAADV
   OUT CP,NORM
END SUB

SUB READMEM (DVALUE)   'Steps 2,3,4 in Sec. 5.3.4
   OUT CP,NORM
   OUT CP,MREAD
   DVALUE = INP(DP)
   OUT CP,NORM
END SUB

SUB WRITEMEM (DVALUE)  'Step 2 in Sec. 5.3.3
   OUT CP,NORM
   OUT DP,DVALUE
   OUT CP,MWRITE
   OUT CP,NORM
END SUB

SUB ENDMEMWR           'Step 7 in Sec. 5.3.3
   OUT CP,NORM
   OUT CP,MENDW
   OUT CP,NORM
END SUB
```

For fast computers, it might be necessary to extend MREAD and MWRITE as previously discussed.

5.4.2 An Example Program Which Uses the Subroutines

The following program tests the memory circuit and illustrates calling the subroutines.

```
'       Program TESTMEMORY
        DEFINT A-Z
        DIM MEMVAL(8192)
        COMMON SHARED CP, DP
        COMMON SHARED NORM, MAADV, MALOAD
        COMMON SHARED MWRITE, MENDW, MREAD
        CP = &H300
        DP = &H301
        NORM = &HFF
        MAADV = &H7F : MALOAD = &HBF
        MREAD = &HFB
        MWRITE = &HFD : MENDW = &HF7
        INPUT "Enter the number of addresses "; NBYTES
        INPUT "Enter the first value to be stored "; SVALUE
        CALL ZEROADR
        FOR I = 0 TO NBYTES-1
          CALL WRITEMEM(SVALUE+I)
          CALL ADVADR
        NEXT I
        CALL ENDMEMWR
        INPUT "Writing Done.  Any Key to Continue "; A$
        CALL ZEROADR
        FOR I = 0 TO NBYTES-1
          CALL READMEM(MEMVAL(I))
          CALL ADVADR
        NEXT I
        INPUT "Reading Done. Any Key to See Values "; A$
        FOR I = 0 TO NBYTES-1
          PRINT MEMVAL(I),
        NEXT I
        END

        Text of  ZEROADR, ADVADR, READMEM, WRITEMEM
        and ENDMEMWR Subroutines
```

The program is minimal. An interesting variation is to input a starting address and then advance to that address before writing and reading. A true test program compares the value written with the value read and lists differences. Note that when SVALUE+I exceeds 255, only the low byte is sent to the memory. So the value read will not equal SVALUE+I. This is solved by comparing the value read with the low 8 bits of SVALUE+I. For example:

```
IF MEMVAL(I)<>((SVALUE+I) AND 255) THEN GOTO ERROR
```

5.5 Troubleshooting the Interface and Software

Construct on one or two breadboards the memory circuit in Figure 5.6. Refer to pin assignments in Appendix C and pin numbers in the figure. Carefully wire all +5's and grounds (power supply connections are not shown in the figure). Enter, save and print the program in Section 5.4. Don't forget the five subroutines.

Check all +5 and ground connections with a logic probe. Proof the program and correct errors. Be especially thorough with constant names and values. Run the program. Despite all precautions, don't expect the system to work. It's likely that something is wrong with the software, parallel ports or hardware. The following is a reasonable troubleshooting procedure.

Recall that an infinite QuickBASIC loop is exited by typing CONTROL-BREAK. Also, a program is aborted by typing CONTROL-C in response to an input.

5.5.1 Test the Parallel Ports

Use software in Chapter 3 to check the parallel ports, including cables. Connect input bits to +5 and ground as necessary. Check outputs with a logic probe. Disconnect incorrect bits from the memory circuit and check again. It's essential to know the ports are working.

5.5.2 Test the Address Generator Hardware and Software

Two types of tests are used. In one, subroutines are put in infinite loops so signals can be observed with an oscilloscope. In the other, INPUT statements are added to stop program execution so signals can be checked with a logic probe.

Test Zero Address

Remove program lines from the first INPUT through the last NEXT I. Add the statements:

```
REPT:  CALL ZEROADR
       GOTO REPT
```

Modify the ZEROADR subroutine as follows:

```
       SUB ZEROADR
         OUT CP,NORM
         OUT CP,MALOAD
```

```
INPUT A$              'stop execution here
OUT CP,NORM
OUT CP,MAADV
INPUT A$              'stop execution
OUT CP,NORM
INPUT A$              'stop execution
END SUB
```

Run the modified program. When execution stops the first time, LD* (CO$_6$) should be low. Check with the logic probe directly on pin 11 of all four counters. If LD* is wrong, disconnect the signal from the counters and check again. If now correct, successively reconnect the counters and replace the offending chip. If LD* is still wrong after disconnection, the software is not working (assuming the parallel ports and cables are). Check the constant names and values. A way to be sure values are properly transferred to subroutines is to add PRINT statements in the routines. Do so if necessary.

If LD* is correct, check that all counter outputs are high. If not, be sure the data inputs are high. Check that the CLears are low. If LD* is correct and any counter output is wrong, disconnect the output from the memory chip and check again. Replace chips if necessary.

Once LD* works, push the enter key to continue subroutine execution to the next INPUT. At this point, CU (CO$_7$) should be low. Check with the logic probe on the low counter's pin 5. If incorrect, disconnect the signal and check again. If correct after disconnection, replace the low counter. If incorrect after disconnection, the software is wrong. Verify constant values and correct as necessary.

When CU works, push enter again to advance to the third INPUT statement. At this point, CU should be high and all counter outputs low. If only the low counter's outputs are correct, check the CA to CU on each higher counter.

When LD* and CU check out, remove the INPUT statements from ZEROADR and restart the program. An infinite loop is executed allowing LD* and CU to be observed with an oscilloscope. Connect LD* to channel 1 and CU to channel 2. With the scope triggering on channel 1, check that first LD* and then CU briefly activate. Look for noise on both lines and, if necessary, add .01 μf capacitors between the controls and ground.

Test Address Advance

Change the modified program by replacing CALL ZEROADR with CALL ADVADR. Do not add input statements. Run the program. Connect CU to oscilloscope channel 1. Observe starting with the lowest bit the counter outputs on channel 2. The low bit should toggle every time CU goes high. The next bit should toggle every other time. Continue for as many bits as practical. Check higher bits with a logic probe. If CU worked in the zero address test above, the likely hardware problem here is noise which might inadvertently load the counters or cause unwanted increments. If necessary, add a .01 μf capacitor between CU and ground and observe the effect.

A static test is to add an INPUT A$ between the CALL and GOTO in the main program. Run the program. When execution stops, use a logic probe to see if the address has advanced 1 from the previous stop. Use this tedious procedure only if the dynamic test above is ambiguous.

5.5.3 Test Memory Write and End Memory Write

Once the address generator hardware and controlling software work, the memory write operation is tested as follows. Modify the program in the last procedure by replacing CALL ADVADR with CALL WRITEMEM(0). Add an INPUT A$ statement in the subroutine after OUT CP,MWRITE. Run the program. When execution stops for the input, check the following:

(1)   WE* (CO_1) should be low at the memory chip's pen 27 and the flip-flop's CLEAR*. If not, check the 7474 and 6264. If necessary, check constant values in the subroutine.
(2)   The 74373's OC* should be active. If not, check the flip-flop and 74373 chip. Be sure the 7474's PRESET* and DATA connections are high.
(3)   All values on $d_7...d_0$ should be low (since zero was output to the data port). Check on the memory's data pins with the logic probe. If any bit is wrong, check the 74373, the 6264 and the values of DP and DVALUE in the subroutine.

A good idea here is to change the program to write 1 and check that d_0 is high. Modify again to write 2 and check d_1. Repeat with 4, 8, 16, 32, 64, 128 and check the other bits.

Next, remove the input statement from the write memory subroutine. After CALL WRITEMEM(128), add the statement CALL ENDMEMWR. Execute the program. Connect WE* to scope channel 1 and EW (CO_3) to channel 2. Trigger on channel 1. The pattern should show WE* activation followed by EW activation. If not, check the flip-flop and software constants. If both signals are as expected, put the 74373 ON/OFF (OC*) on channel 2. WE* should activate OC*, and the end of EW should deactivate the signal. If not, noise could be inadvertently setting and/or clearing the flip-flop.

According to specifications in Section 5.1, WE* must be active at least 90 ns. Measure WE*'s length on the scope and add outputs if it is not 250 ns.

Finally, check that the 6264's CE_1* (pin 20) is low and CE_2 (pin 26) is high.

5.5.4 Test Memory Read

The final test is read memory. Again, modify the program in the last procedure by replacing the two calls with CALL READMEM(X). Add an INPUT

A\$ statement to the subroutine after OUT CP,MREAD. Run the program. When execution stops, check at memory pin 22 that OE* (CO_2) is low. If not, disconnect the signal from the memory chip and check again. If OC* is still wrong, check software constant values. If OE* is correct, use the logic probe to see that each memory data pin is high or low. If not, the memory chip is defective. (The actual values are meaningless).

Remove the INPUT statement from the subroutine. Again, run the program. Connect OE* to oscilloscope channel 1 and verify the signal's activation as the subroutine is repeatedly called. Next, connect the flip-flop's output (OC*) to channel 2. It must always be high. There is a possibility that noise from OE*'s activation inadvertently clears the flip-flop. Add noise capacitors if necessary.

According to specifications in Section 5.1, OE* must be active at least 70 ns before data is read. Measure OE*'s length on the scope and add outputs if it is not, say, 500 ns.

5.5.5 Summary

After all the tests above are positive, load the original program. Correct the software errors that were found.

> *A problem is forgetting to correct the original program or ending up with several copies with some corrected and others not.*

If the entire system still doesn't work, the tests above could be repeated. Another approach is to immediately put input statements in the subroutines and check the five control signals as the main program runs. Look for clues in how the system fails. For instance, if the same value is read from all memory addresses, the address generator may not be incrementing or the 74373's OC* is on during reads.

Troubleshooting hardware/software systems requires systematically determining what works in order to identify what doesn't. Wiring errors and omissions are common as are misspelled constant names and incorrect constant values. Defective chips are rare.

5.6 Example Application - A Digital Recorder

Memory is useful in two categories of measurement and control systems: fast data recorders and high frequency signal generators.

In recorder systems, hardware stores values at successive addresses at a fixed rate (every Δt seconds). After recording, software reads the memory and analyzes the data using the fact that the bytes were recorded Δt seconds apart. With the 6264, saved values may be 8-bit binary data from devices such as ADC's or they may be up to eight individual status and control signals.

In signal generator systems, a predetermined sequence of bytes is written by software to successive memory addresses. Then, after initiation, the hardware reads the bytes back from successive addresses at a fixed rate. The values serve either as 8-bit data for devices such as digital-to-analog converters (DAC's) or as individual timing and control signals. The frequency depends on the rate of reading. Only part or the entire memory capacity can be used, and the output sequence may occur when triggered or be continually repeated.

The circuit and software for an 8-channel digital signal recorder are presented here as an example memory application. The instrument's specifications are:

(1)   Up to 8 digital signals may be recorded. They are connected to the 8 data lines of a 6264 memory.

(2)   Recording is initiated by the host computer.

(3)   After initiation, the system automatically stores the high or low state of each signal every microsecond for 8191 μsec. After recording, address 0 contains the initial or time = 0 state of the 8 signals. Address 1 contains the state 1.0000 μsec later. Address 2 contains the state after 2.0000 μsec. Address 8191 contains the state after 8191.0 μsec.

To illustrate what the recorded data mean, suppose these values are found at addresses 0 to 11.

address	binary value
0	11101101
1	11101101
2	11011101
3	10011101
4	10001001
5	00001100
6	00111100
7	00111100
8	01101100
9	11101000
10	11011000
11	11011000

Figure 5.7 is a graph of the signals verses time. The "diagonal" transitions indicate that the signal reverses state sometime between one recording and the next.

The recorder example is intended to show how to design and operate a specific system using the parallel ports and a 6264 memory. Several fundamental logic design ideas are illustrated. A context is provided for interesting problems such as how to add signal capacity (to 16 or 32 bits) and how to add storage capacity (to 16K or 32K bytes). Further, the recorder provides an illustration of the utility of programmable interval timers or PIT's. Finally,

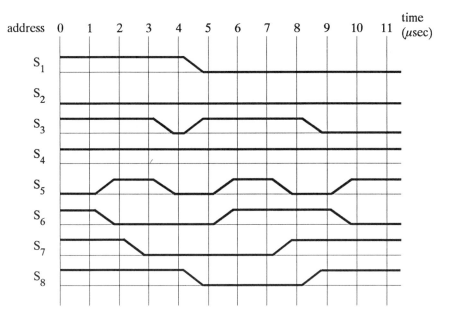

Figure 5.7. Time plot of values stored in memory which shows each signal's state the first 11 μsec of recording.

the example should provide a background sufficient to design other data recorder and signal generator systems.

5.6.1 Recorder Design Principles

The first issue is: how can hardware replace software in storing values at successive 6264 addresses? Assume the following setup:

(1) The address generator described in Section 5.2 supplies $A_{12}...A_0$ to a 6264. Recall that the signal LD* makes $A_{12}...A_0$ all 1's, and the signal CU going low-high increments the address.

(2) An accurate, symmetric 1.0000 MHz digital clock is available. Symmetric means high for half a cycle and low the other half.

(3) Up to 8 digital signals are connected to the 6264 data lines.

(4) The 6264 CE's are wired active and its OE* (used for reading) is inactive.

(5) Activation of LD* has made the address lines all high.

Figure 5.8 shows a scheme for hardware memory storage. At some point (to be worked out), the 1.0000 MHz clock and its inverse are passed to CU and to the 6264's WE*, respectively. Because of step 5 above, the first time CU goes high, the address rolls over to 0. Then 1/2 clock cycle later, WE* goes from low to high storing the current signal states at address 0. 1/2 cycle

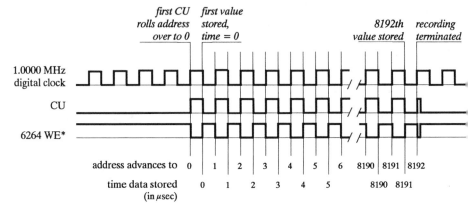

Figure 5.8. Scheme for hardware to store signal states at successive 6264 addresses.

later, CU goes high again making the address 1. 1/2 cycle after that, WE* goes high storing the signal states at address 1. The process continues until the signal states 8191 μsec after the start are stored at address 8191. Then when the address advances to 8192, data acquisition stops (how to be worked out).

It's important to be sure the scheme's timing is consistent with 6264 specifications. As shown in Figure 5.3, the address must be stable t_{AW} seconds before WE* goes high. With a 1 MHz clock, a cycle is 1 μsec long and half a cycle is 500 ns. Therefore, t_{AW} is 500 ns, more than enough. Also, data must be stable t_{DW} seconds before WE*. A typical minimum value is 60 ns. So if a signal changes within 60 ns of WE* going high, its previous value may be recorded.

Once a basic recording scheme is determined, the problems become:

(1) How can software initiate recording?

(2) How can hardware terminate recording immediately after address 8191 is written?

(3) How can the host computer later read the stored values?

5.6.2 Recorder Initiation

The objective is to initiate the recording sequence in Figure 5.8. Suppose control-out bit 3 is chosen to originate a software START signal. The circuit in Figure 5.9 accomplishes the goal and works as follows.

(1) At the beginning of the recorder program, an output to the control port makes START (CO_3) low. This clears the 7474 flip-flop, and the signal labeled S-START (for Software **START**) is low. Since one input to the AND gate is low, its output is low and CLOCK is not passed on.

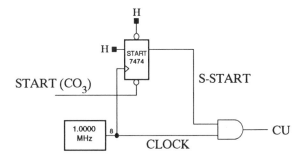

Figure 5.9. Circuitry which uses CO_3 to initiate the recording sequence.

(2) After making START low, the recorder program carries out other ini-
     tializations and then enables the hardware by putting START high.

(3) START may go high at any point in a hardware clock cycle. The action
     has no effect on the flip-flop. However, the next time CLOCK goes high,
     the FF's output S-START goes high (via the clock-data mode). Now,
     with one of the AND gate's inputs high, the output equals the other input.
     Hence, CLOCK is passed on and the scheme in Figure 5.8 begins.

5.6.3 Recorder Termination

Recording must stop when values have been stored in all 8192 addresses.
Because of the sequence's speed, this must be done in hardware. The question is how? What signal changes at the right point? The address generating
counter which supplies A_{12} has three unused outputs. Consider what the next
higher bit after A_{12} becomes when CU goes high with the address already 8191.

	address
decimal	**binary counter outputs**
8190	0001 1111 1111 1110
8191	0001 1111 1111 1111
8192	0010 0000 0000 0000

The fourth counter's QB (what would be A_{13}) uniquely goes high at 8192. The
signal stops data recording as shown in Figure 5.10.

(1) START is initially low which makes S-START low. This in turn makes
     the inverse Q* output of the stop flip-flop high and the signal H-STOP*
     (for **H**ardware **STOP**) high.

(2) When START and then S-START go high, H-STOP* is unaffected and
     remains high. Hence CLOCK passes through both AND gates produc-
     ing H-CU (Hardware Count-Up) and the 6264's WE*, as specified ear-
     lier.

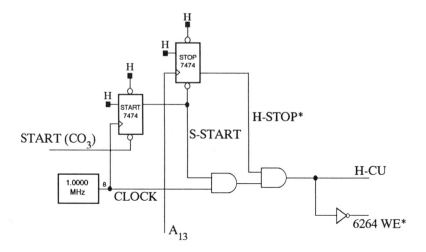

Figure 5.10. Circuitry which uses A_{13} to terminate hardware recording.

(3)   The only way H-STOP* can go low is for the signal connected to the stop
       FF's clock input to go high. This occurs when address 8192 is reached.

       Two flip-flops and two AND gates are a straightforward approach to
initiating and terminating a process. The idea is for flip-flops to generate
"control" inputs to AND gates which must both be high for subsequent circuitry to work. Virtually any set of conditions can set and clear the flip-flops.

5.6.4 Recorder Circuit and Timing

So far, circuit parts which record data in memory have been presented. Figure 5.11 is the complete diagram. Elements not previously discussed enable
software to read values after hardware storage. They are:

(1)   After making START high, the program must wait for data recording
       to end. This is done by connecting H-STOP to control-in bit 0.
(2)   If the signals to be recorded were connected directly to the 6264 data
       pins, they would interfere with software reading stored values during which
       the memory puts values on $d_7...d_0$. Therefore, the signals are buffered
       through a 74373 whose OC* is low only during hardware recording. Since
       S-START is high for this time, the start FF's inverse Q* is low and sup-
       plies OC*.
(3)   Software controlled S-CU (CO_7) is used to advance the address during
       read operations. However, during hardware recording, H-CU is needed.
       (Earlier both signals were simply CU. Now they are distinguished.) In
       the final circuit, S-CU and H-CU are ORed with the result going to the
       low counter. During recording, if S-CU is low, then H-CU is passed to
       the low counter. During software reading, if H-CU is low, S-CU is passed.

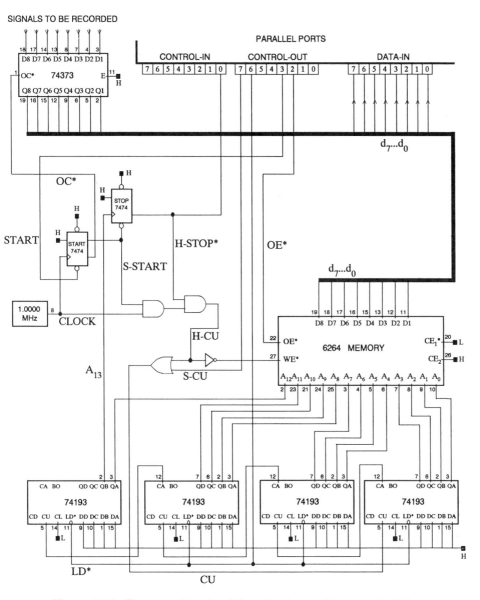

Figure 5.11. The complete circuit for a hardware data recorder interfaced to the parallel ports.

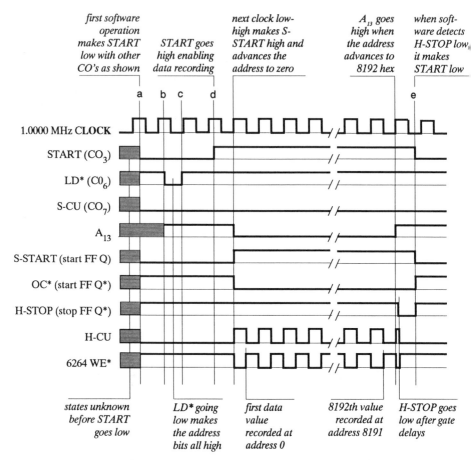

Figure 5.12. Annotated timing diagram for hardware data recorder. The letters just above CLOCK refer to software steps.

(4)   LD* and OE* are unaffected by circuit additions.

The timing diagram in Figure 5.12 summarizes the circuit and its operation.   Highlights are:

(1)   Software makes START low (step a in Figure 5.12).
(2)   Software briefly activates LD* making all the address bits 1 (steps b and c).
(3)   Software puts START high (step d).
(4)   The next CLOCK low-high makes S-START high (H-STOP is initially high). H-CU and WE* then carry out hardware recording.
(5)   When the address reaches 8192, A_{13} goes high. H-STOP goes low and hardware recording stops.

(6) Meanwhile, software monitors H-STOP (at CI_0). When detected low, START is again made low and the circuit returns to its initial state (point e in Figure 5.12).

(a) Note that S-CU is low during the entire record sequence allowing H-CU to pass through the OR gate.

(b) Note the gate delay spikes (just after the address advances to 8192) in H-CU and 6264 WE*. The latter might cause erroneous data by rerecording a value in address zero. The easy thing to do is ignore the address.

After making START low, the previous ZEROADR, READMEM and ADVADR routines input the saved values.

5.6.5 Software

In formulating software which works through the parallel ports, constants are envisioned which when sequentially sent to control-out causes hardware to produce the desired timing pattern. For the recorder system, the normal, inactive state of the control-out bits is:

CO_7	CO_6	CO_5	CO_4	CO_3	CO_2	CO_1	CO_0
S-CU	LD*	X	X	ST	OE*	X	X
1	1	1	1	0	1	1	1

Note that control-out bit 3 is low in the inactive state. This exception to the rule of high inactive states is necessary so H-CU is passed to the address counters during hardware record. The recorder requires the following four software operations.

(1) Start hardware recording with the following sequence which carries out steps a, b, c and d in Figure 5.12.

CO_7	CO_6	CO_5	CO_4	CO_3	CO_2	CO_1	CO_0	hex value	constant name
S-CU	LD*	X	X	ST	OE*	X	X		
1	1	1	1	0	1	1	1	F7	NORM
1	0	1	1	0	1	1	1	B7	MALOAD
1	1	1	1	0	1	1	1	F7	NORM
0	1	1	1	1	1	1	1	7F	MSTART

(2) Poll CI_0 awaiting the end of recording when H-STOP goes low.

(3) After termination, output NORM to the control port (step e in Figure 5.12).

(4) Read stored values by calling the ZEROADR, READMEM and AD-VADR subroutines with the following constants (which differ from earlier values only in keeping CO_3 low).

CO_7	CO_6	CO_5	CO_4	CO_3	CO_2	CO_1	CO_0	hex	constant
S-CU	LD*	X	X	ST	OE*	X	X	value	name
1	1	1	1	0	1	1	1	F7	NORM
1	0	1	1	0	1	1	1	B7	MALOAD
0	1	1	1	0	1	1	1	77	MAADV
1	1	1	1	0	0	1	1	F3	MREAD

The constants NORM, MALOAD, MAADV and MREAD are different from when software stored values in memory. So it might be advisable to use different constant names. However, in Microsoft QuickBASIC (and many other languages), constants (and variables) are local to subroutines unless declared shared. One reason for the practice of main programs assigning values to constants and declaring them shared is that the same routines can be used with different constants.

5.6.6 Example Program

The following program starts hardware recording, waits for termination, reads the first 500 stored values and prints "0" if bit 0 (Signal 0) is low and "1" if bit 0 is high.

As in other examples, the program demonstrates every detail of interacting with the parallel ports and digital recorder circuit. Clearly, far more elaborate and flexible software with graphics is desirable.

```
'         Program RECORDER
          DEFINT A-Z
          DIM RECVALUE(500)
          COMMON SHARED CP,DP,NORM
          COMMON SHARED MAADV,MALOAD,MREAD
          CP = &H300 : DP = &H301
          NORM = &HF7
          MAADV = &H77 : MALOAD = &HB7
          MREAD = &HF3
          MSTART = &H7F
          CLS
          INPUT "Any key to start hardware recording ";A$
          OUT CP,NORM
          OUT CP,MALOAD
          OUT CP,NORM
          OUT CP,MSTART
WWH:      IF (INP(CP) AND 1) = 1 THEN GOTO WWH
          PRINT "Recording Complete, Reading Underway"
          CALL ZEROADR
          FOR I = 0 TO 499
            CALL READMEM(RECVALUE(I))
            CALL ADVADR
```

```
NEXT I
CLS
PRINT "Reading Complete, Values of Bit 0"
FOR I = 0 TO 499
  IF (RECVALUE(I) AND 1) = 1 PRINT "1";
      ELSE PRINT "0";
NEXT I
END
```

Text of ZEROADR, ADVADR, and READMEM subroutines

5.6.7 Recorder Limitations

The recorder hardware and software demonstrate important design and control techniques. However, the system is not a reasonable instrument. Among others, the following limitations are serious:

(1) The rate of recording signal states is fixed at 1.0 MHz. In a practical system, the rate should be programmable.

(2) The number of readings is fixed at 8192 but should also be programmable.

(3) The initiation of recording is controlled by software and is not coordinated with what's being measured. There should be a hardware trigger.

Solution of the first two limitations requires programmable interval timers or PIT's, the subject of the next chapter. An exercise (at the end of the chapter) involves hardware triggering.

5.7 Troubleshooting the Recorder

Construct on two breadboards the recorder circuit in Figure 5.11. Refer to pin assignments in Appendix C and pin numbers in the figure. Carefully wire all +5's and grounds (power supply connections are not shown in the figure). Enter, save and print the program in Section 5.6.6. Don't forget the three subroutines.

Check all +5 and ground connections with a logic probe. Proof the program and correct errors. Run the program. Despite all precautions, don't expect the system to work. It's likely that something is wrong with the software, parallel ports or hardware. The following is a reasonable troubleshooting procedure.

Recall that an infinite QuickBASIC loop is exited by typing CONTROL-BREAK. Also, a program is aborted by typing CONTROL-C in response to an input.

5.7.1 Test the Parallel Ports

If not carried out earlier, use software in Chapter 3 to check the parallel ports, including cables. Connect input bits to +5 and ground as necessary. Check outputs with a logic probe. Disconnect incorrect bits from the recorder circuit and check again. It's essential to know the ports are working.

5.7.2 Test the Address Generator Hardware and Software

Perform the address generator tests outlined in Section 5.5.2. For CU to work as described, S-CU must pass the OR gate. Disconnect H-CU from the AND gate and wire low. Modify the program by removing lines from CLS to the last NEXT I.

5.7.3 Test Memory Read

Follow the procedure in Section 5.5.4 to test memory read. Again, wire H-CU low.

5.7.4 Test Start and Hardware Recording

After the address generator and memory read work, reconnect H-CU to the AND gate. Load the main program. Add INPUT A$ statements just before and just after OUT CP,MSTART. Remove A_{13} from the stop flip-flop. Run the program. When execution stops for the first input, check the following signals with the logic probe directly on chip pins.
(1)   START (CO_3) should be low. If not, check software constant values.
(2)   S-START should also be low. If START is low and S-START is wrong, check wiring, both flip-flops and the AND gate.
(3)   The 74373's OC* should be high.
(4)   H-STOP (also CI_0) should be high.
(5)   H-CU should be low and the memory's WE* high.
(6)   OE* (CO_2) should be high.
(7)   All address bits should be high. Check directly on memory pins.
     When all these signals are correct, push the enter key to continue program execution to the second added INPUT. Check the following signals with the logic probe.
(1)   START (CO_3) should be high. If not, check the value of MSTART.
(2)   S-START should also be high. If START is high and S-START is wrong, check CLOCK with an oscilloscope.
(3)   The 74373's OC* should be low. Wire 74373 input bit D1 low (leaving the rest unconnected). Check the outputs at the memory's data pins. Recall that disconnected inputs are seen as high.
(4)   H-STOP should still be high.
     Because A_{13} was removed from the stop flip-flop, hardware recording should be continuous allowing signals to be observed on an oscilloscope. With CLOCK connected to channel 1, observe the following on channel 2.
(1)   H-C and CU should be the same as CLOCK.

(2) WE* should be the inverse of CLOCK.
(3) The address bits should be continually cycling. The low bit should be the toggle of CLOCK; the next should toggle every other CLOCK; etc.
(4) A_{13} should be mostly low but occasionally go high.
(5) Be sure there is no noise on the memory's OE* control.

When all these signals work as expected, abort the program and reconnect A_{13}. Reload and run the main program. The system should work. If not, repeat the tests above. Look for clues based on what actually happens.

5.8 Summary

Digital design with memory integrated circuits involves read and write timing sequences, management of data lines and address generation. In all cases, obtain manufacturers' specifications and master the timing diagrams. After interfacing memory to the parallel ports, the next task is designing "fast" hardware which writes values to memory, reads previously stored values from memory or some combination. The recorder example provides a strategy applicable to a variety of cases.

Exercises

1. Use parallel ports (which meet the specifications in Chapter 3) and a breadboard system to build the memory circuit in Figure 5.6. IC pin assignments are in Appendix C. Enter the test program in Section 5.4. Carry out the troubleshooting sequence in Section 5.5.

2. Write a program which asks for an address and value and then stores the value at the address. Write a second program which asks for an address and then displays the value previously stored at that address.

3. Design a 32K by 8-bit memory circuit consisting of four 6264's interfaced to the parallel ports.

4. How could one make an 8K by 16-bit memory? The 8 data lines from the two 6264's must be multiplexed. Recall that 3 control-out bits are unused.

5. Build, test and troubleshoot the digital recorder described in Sections 5.6 and 5.7.

6. Design a hardware trigger for the digital recorder. Specifically, make recording start the first time the bit 0 signal goes high after a software start.

7. How could the digital recorder program be modified to compute the frequency of one of the signals recorded? What range of frequencies (at what accuracy) might be determined?

8. Design a circuit and write software for a digital signal generator. Specifically, software downloads 8192 predetermined values to a 6264. Then, under the control of a 1.0 MHz clock, hardware sequentially reads and makes available the stored values.

9. An important instrument in studying radioactive decay is the pulse height analyzer. Assume an ADC produces a 10-bit binary value proportional to the energy of detected radiation. The ADC outputs supply the address $(A_9...A_0)$ to a 6264 and, following each detection, one is added to the value stored at the address. After many detections and additions, each memory location (or channel) contains the number of times radiation of a particular energy was detected.

    Devise a hardware scheme to accomplish pulse height analysis. An approach is to load the addressed memory value into two 4-bit counters, count up one, and then store the incremented value back at the same address. A couple 74373 latch/buffers are needed to coordinate the data lines. The most difficult problem is generating controls signals following each detection (such as 6264 WE* and OE*, 74373 OC*'s, and counter LOAD* and CU). Determine the overall circuit and control sequence but leave actual control signal generation until after PIT's are mastered. Don't forget that software must initially zero the memory and read data after acquisition.

    While an interesting exercise, this is not a practical instrument unless each address can hold more than 8-bit values. To achieve reasonable 24-bit values, three 6264's are needed as well as more buffers, counters and controls. A plausible but slow "software" pulse height analyzer is described in Section 9.4.6.

6 Tutorial on the 8253 PIT

The Intel 8253 programmable interval timer (PIT) is the third and most important building block for measurement circuitry. Once the chip is mastered and interfaced to the parallel ports, the door opens to a limitless assortment of timing and counting applications. The next several sections describe first how the 8253 may be used to generate timing and control signals and then how it may be used to acquire data by counting events and periods. Next, programming, loading and reading the device are described in detail. How to interface the PIT to parallel ports and control its operation with software are given. A complete example application (a manually triggered scaler) is presented. The chapter concludes with a discussion of a potential design flaw and its solution.

6.1 Basic Operations

The 8253 is a 24-pin integrated circuit. It contains three independent 16-bit counters (designated CTR_0, CTR_1, CTR_2). As diagramed in Figure 6.1, each counter has two 16-bit registers (storage and count) and each has clock and gate inputs and an output.

Once interfaced to the parallel ports, each counter may be programmed to operate in any one of six modes, numbered 0 to 5. Each storage register

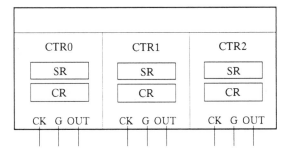

Figure 6.1. Logic drawing of an 8253 PIT showing the three independent counters. Each has a storage register (SR) and a count register (CR) and each has clock and gate inputs and an output.

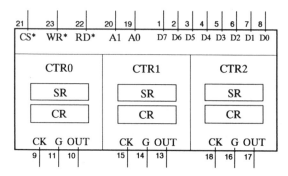

Figure 6.2. Complete logic drawing of the 8253 showing the control and data connections. A separate pin assignment drawing is in Appendix C.

may be loaded by software with any value between 0000 and FFFF hex. After programming and loading, depending on the mode, actions at the clock and gate inputs cause the contents of the storage register to be copied to the count register. Then, no matter what mode, while gate is high every clock high-low transition causes the count to decrement by 1. The value of the count can be read at any time. The pattern followed by the output as well as what happens when the count reaches zero depend on the mode.

In addition to clock, gate and output connections (9 in all), an 8253 has 8 pins ($D_7...D_0$) for bi-directional **D**ata transfers and the following five control inputs: CS* (Chip **S**elect), WR* (**WR**ite), RD* (**R**ea**D**), A_0 (Address **0**) and A_1 (Address **1**). All logic connections are shown in Figure 6.2. Interfacing the 5 control and 8 data lines to the parallel ports enables software to program modes, load storage registers and read count values. Exactly how to carry out these operations is described in Section 6.4. But first, there are two basic types of applications to consider: those where the goal is to generate specific output patterns for timing and control purposes, and those where the goal is to count clock transitions and afterward read the count value.

6.2 Output Pattern Generation

Very accurate clock integrated circuits were discussed in Section 2.3. Their outputs are symmetric square waves (high for half a cycle and low the other half). Typical frequencies are 1 and 2 MHz. Such master clocks may be used with PIT counters to produce a variety of timing and control signals.

6.2.1 Mode 1: Programmable One-shot

Digital pulses are needed for operations such as initiating an analog-to-digital conversion or determining the length of a counting interval. The 8253's mode 1 generates triggerable, programmable length pulses.

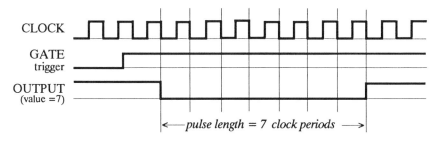

Figure 6.3. Output pattern for mode 1 (one-shot) with the loaded value = 7.

Suppose an accurate frequency master clock is connected to the clock input of a PIT counter, mode 1 is programmed and the storage register is loaded with a value. Upon trigger the counter's output goes low for a length of time equal to the loaded value times the period of the clock input. Before the trigger, the output is high and it returns high at the end of the interval. The pattern in Figure 6.3 is for value = 7.

The range of pulse lengths depends on the frequency of the clock input. If it is 1 MHz (period 1 μsec), the shortest pulse would be 1 μsec (value = 1), and the longest would be 65535 μsec (value = 65535, the largest possible 16-bit number).

Triggering is accomplished by a low-high transition at the gate input. After a pulse is over, another equal length pulse may be triggered by another low-high at the gate input.

The exact operation is as follows. The end of the first low-high-low clock sequence after gate goes high causes: 1) the contents of the storage register (value) to be copied into the count register; and 2) the output to go low. Then every subsequent clock high-low decrements the count by 1. When the count reaches zero, output returns high and the pulse stops.

Pulses end with the count register decremented to zero. However, the separate storage register contains the original value loaded by software. This enables another equal length pulse to be triggered without any action by software, something which is often quite useful.

The seemingly tedious fact that the count register is not loaded with value (from the storage register) until the first low-high-low clock sequence after trigger is actually quite important. The problem is that trigger may occur at any time during a clock cycle. Yet for the pulse to be an integral number of clock cycles long, it must be synchronized with the clock. Therefore, it's necessary for the pulse to be held up until the beginning of the next full clock cycle. Figure 6.4 highlights synchronization, again for the case of value = 7.

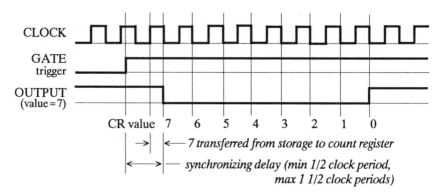

Figure 6.4. Illustration of mode 1 synchronization.

6.2.2 Modes 2 and 3: Rate and Clock Generators

Unlike mode 1 where a pulse is triggered by the gate, the outputs in modes 2 and 3 are continuous periodic signals as shown in Figure 6.5.

A common goal is to create from a master additional clock signals with lower frequencies. PIT modes 2 and 3 accomplish this. Specifically, the master clock is connected to the clock input of a counter, the counter is programmed for mode 2 or 3, and its storage register is loaded with a 16-bit value. If the gate input is high or when it goes high, the next low-high-low clock sequence causes the contents of the storage register to be transferred to the count register. Beginning at this point, the counter's output is a continuous signal with a period equal to the clock period times the loaded value. To change the output period, just load a different value into the storage register.

Suppose the input is a 1 MHz master clock (period of 10^{-6} second); then the outputs for various loaded values are:

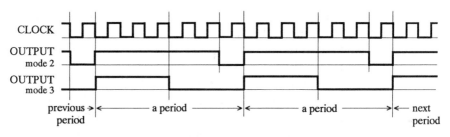

Figure 6.5. Output pattern for modes 2 and 3 (rate and clock generators) with the loaded value 6 (in both modes). Gate (not shown) is high.

loaded value	output period	output frequency
5	5×10^{-6} sec	200 KHz
468	468×10^{-6} sec	2.136 KHz
9000	9000×10^{-6} sec	111.1 Hz
25000	25000×10^{-6} sec	40.0 Hz
65535	65535×10^{-6} sec	15.3 Hz

Since 65535 is the largest number which can be loaded into the 16-bit count register, the way to obtain even lower frequencies (longer periods) is to use two PIT counters in succession (with the output of the first connected to the clock input of the second). Starting with 1 MHz, the longest period would be $65535 \times (65535 \times 10^{-6}) = 4295$ seconds (1.19 hours) which corresponds to .000233 Hz. Of course, one could go on to three counters (all in just one 8253) and get even longer periods.

The difference between modes 2 and 3 is that in mode 2 an output cycle is high for all but the final master clock period which makes up the cycle. The signal in mode 3 is a symmetric square wave. Both patterns are shown in Figure 6.5.

The exact operation of mode 2 is that after gate goes high, the first low-high-low clock transition transfers the value previously loaded in the storage register to the count register. The register is decremented by 1 on each subsequent clock high-low. When the count reaches 1, output goes low. When 0 is reached, output returns high and the original value is automatically and instantaneously transferred from the storage to the count register and the

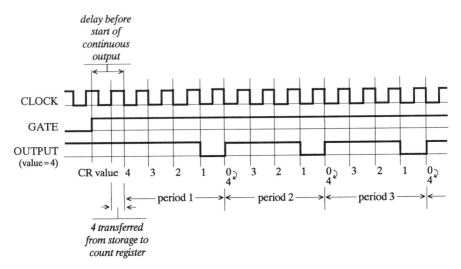

Figure 6.6. Synchronization and count register contents in mode 2 operation. Loaded value = 4.

pattern repeats. Figure 6.6 shows the contents of the count register during the first three periods of mode 2 operation.

The effect of gate going low is to stop the decrementing process. When gate returns high, the next clock low-high-low reloads the storage register value into the count register and restarts the output pattern.

The exact operation of mode 3 is that once the contents of the storage register are transferred to the count register, each clock high-low decrements the count by 2. When the count reaches zero, the output reverses state and the storage register's contents are automatically and instantaneously transferred to the count register and the process repeats (Figure 6.7). Hence, the output is a symmetric square wave with each half cycle VALUE/2 clock periods long. If the loaded value is an odd number, whenever the output reverses to high, the first decrement is 1 (and thereafter 2); and when the output reverses to low, the first decrement is 3 (and thereafter 2). Hence, when value is odd, the high half-cycle is one input clock period longer than the low half-cycle. Figure 6.7 shows the contents of the count register during the first two periods when value = 5.

6.2.3 Modes 0, 4 and 5

An 8253 PIT can be programmed for any one of six modes. Once programmed, a mode is maintained either until the power is turned off or another mode is

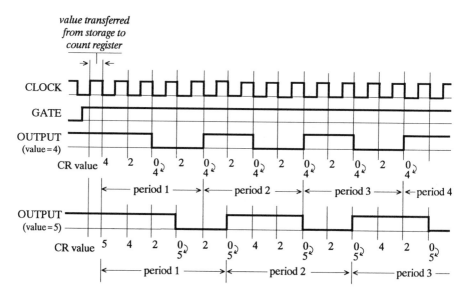

Figure 6.7. Count register contents during mode 3 operation for value even (top) and value odd (bottom).

programmed. The cases outlined above are sufficient for most operations. Once they are mastered, modes 0, 4 and 5 might be considered. They will be described as needed in later chapters, and are outlined in the complete 8253 specifications in Intel's *Peripheral Design Handbook* (reference in Appendix A).

For the purpose of completeness, mode 0, interrupt on terminal count, is similar to the one-shot (mode 1) except the low pulse is triggered by software loading a value into the storage register. Mode 4 (software triggered strobe) and mode 5 (hardware triggered strobe) produce active low outputs one input clock period long after the count decrementes to zero. In mode 4, decrementing starts as soon as a value is loaded into the storage register; and, in mode 5, decrementing starts when the hardware gate goes high.

6.3 Using Counters for Measurement

In addition to creating output patterns (one-shot, rate clock, square-wave clock, and others), PIT counters may also be used to count the number of high-low transitions of the clock input during some time interval (after which the contents of the count register may be read by host software). For instance, time itself is measured by counting the number of cycles of a known frequency reference clock during an interval (with the interval length equaling the number of cycles counted multiplied by the known reference period). The issues are what mode to use and how to control the counting interval with the gate input.

It turns out that any mode could be used, except 3 where the decrement is by 2. However, actions of the gate input in mode 2 make it especially effective. With the gate low, software sends a value to a counter's storage register. When gate goes high, the next clock low-high-low loads the value into the count register. Then every subsequent clock high-low decrements the count by 1. Finally, when gate returns low, counting stops and software reads the contents of the count register. The process is shown in Figure 6.8 for a non-periodic clock signal and value = 100. During the first high gate interval, clock goes high-low four times. But the counter decrements only three times because the first low-high-low loads the storage register value into the count register. Hence when the count is read, the actual number of clock high-low transitions must be computed as:

(loaded value - decremented value read from the count register) + 1

To repeat counting for another gate interval, simply have gate go high again. The next clock low-high-low reloads the count register with the original value and decrementing continues as long as gate remains high.

As stated several times earlier, the storage register value is not loaded into the count register until after gate goes high and subsequently clock goes

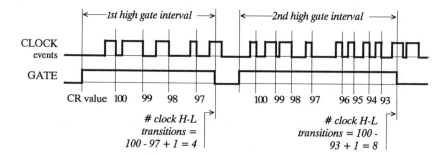

Figure 6.8. Counting measurement using mode 2 with the loaded value = 100. Computation of the number of clock high-low transitions (events) during a high gate interval is computed from the value read from the count register as shown.

low-high-low. This is a blessing when the objective is to synchronize a pulse output with the beginning of the next full clock cycle. But it's a problem here. As indicated above, if one wants to know the total number of clock high-low transitions during an interval, 1 must be added. Further, if the clock signal makes no low-high-low transitions during an interval (which could be the case in some measurement circumstances), no value is loaded into the count register and the read operation yields indeterminate results. In such cases, additional logic must be added to produce an initial low-high-low of the clock signal. This is discussed in the context of an example application in Section 6.8.

Figure 6.9 further demonstrates problems with accuracy. During the gate interval in Case A, there are 10 high-low clock transitions but only 9 are counted (after adding 1). But, in Case B there are 11 high-low transitions and all 11 are counted (after adding 1). The difference in the two cases is a "slightly" longer gate interval in B and "significantly" different synchronizations. Awareness of exactly how counting works is sufficient to avoid potential errors, which are significant only when the anticipated number of clock high-lows during the counting interval is "small."

Finally, in the other extreme when there are thousands of high-low transitions, it's necessary to keep the number of clock transitions less than the loaded value, otherwise subtraction leads to a negative number. In general, the loaded value should be as large as possible up to the maximum of FFFF hex or 65535.

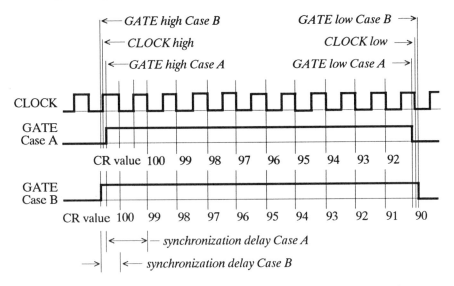

Figure 6.9. Illustration of potential counting errors in two cases with "slightly" different high gate intervals.

6.4 Programming, Loading and Reading

As stated earlier, the 8253 PIT contains three counters. Each may be programmed for one of six modes; the storage register may be loaded with any 16-bit value; and the current contents of the count register may be read by software. Five control lines and eight data connections carry out these operations. This section presents the program, load and read sequences specified by the manufacturer. The next section shows how the 8253 can be interfaced to the parallel ports described in Chapter 3.

6.4.1 PIT Connections

In addition to clock and gate inputs and an output for each of the three counters, an 8253 has five control inputs and eight data connections. They are:

CS*   Chip Select must be active for an 8253 to respond to program, load and read operations

A_1 and A_0 Address lines specify (when CS* is active) which of the three counters is being loaded or read as follows:

A_1	A_0	
0	0	CTR_0 is loaded or read
0	1	CTR_1 is loaded or read
1	0	CTR_2 is loaded or read
1	1	mode is programmed

WR*	WRite must be active when program and load operations occur (when information originating with the host computer is being sent to the PIT)
RD*	ReaD must be active when read operations occur (when information originating with the PIT goes to the host computer)
$D_7...D_0$	8 bi-directional Data lines operate as follows:

when WR* and CS* are active, the 8253 expects host software to put a value on the data lines

when RD* and CS* are active, the 8253 itself puts a value out on the data lines (which host software reads)

when neither WR* and CS* or RD* and CS* are active, $D_7...D_0$ is in a Hi-Z state

6.4.2 Sequence to Program the Mode

The timing diagram in Figure 6.10 shows the six step sequence of the control and data lines needed to program the mode of a PIT counter. As labeled in the figure, the steps are:

(a) This is the normal or inactive state: CS*, RD* and WR* are all inactive; A_1A_0 can be anything (either high or low); and $D_7...D_0$ is in a Hi-Z (disconnected) state.

(b) The first action occurs: CS* goes active and A_1A_0 must both be high.

(c) An 8-bit value is output by the computer. This value, called the Control Word (CW), tells the PIT which of the 3 counters is being programmed, what the mode is, and other information as described below.

(d) WR* goes active. (In the particular interface scheme to be used, it is not until WR* goes active that the byte output in step c is actually put on the data lines.)

(e) WR* returns inactive. This is the point when the PIT latches CW.

(f) Finally, the normal, inactive state is restored. Specifically, CS* returns inactive, A_1A_0 can again be either high or low, and $D_7...D_0$ returns to a Hi-Z state.

Critical times, shown in Figure 6.10, are that WR* must be active at least $t_{WW} = 400$ ns, and data must remain at least $t_{WD} = 40$ ns after WR* returns inactive. (Consult Intel's specifications for additional timing information and for variations among the different speed 8253's.)

The relationship between the 8-bit value of CW ($CW_7...CW_0$) and the action produced is:

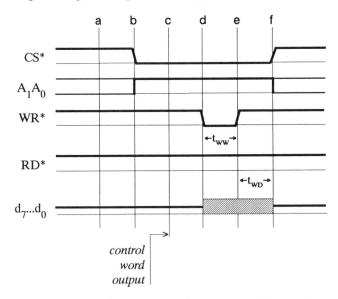

Figure 6.10. Timing sequence for program PIT operation.

CW_7	CW_6	action of the high 2 bits
0	0	program counter zero
0	1	program counter one
1	0	program counter two

CW_5	CW_4	action of bits 5 and 4
1	1	program 16-bit operation of the selected counter

(Options specified by other values of CW_5CW_4 include 8-bit operation. See Intel's specifications.)

CW_3	CW_2	CW_1	mode selection bits
0	0	0	mode 0
0	0	1	mode 1
0	1	0	mode 2
0	1	1	mode 3
1	0	0	mode 4
1	0	1	mode 5

CW_0	action of the low bit
0	16-bit counter is pure binary
1	16-bit counter acts as a four digit decade counter

Example Control Words

CW₇ CW₆ CW₅ CW₄ CW₃ CW₂ CW₁ CW₀

$$CW_7\ CW_6\ \ CW_5\ CW_4\ \ CW_3\ CW_2\ CW_1\ \ CW_0$$

0   0   1   1   0   0   1   0   (50 decimal or 32 hex)

Programs PIT CTR_0 for 16-bit operation, mode 1, with a binary format.

$$CW_7\ CW_6\ \ CW_5\ CW_4\ \ CW_3\ CW_2\ CW_1\ \ CW_0$$

1   0   1   1   0   1   1   0   (182 decimal or B6 hex)

Programs PIT CTR_2 for 16-bit operation, mode 3, with a binary format.

6.4.3 Sequence to Load a 16-bit Value into a Selected PIT Counter

The timing diagram in Figure 6.11 shows the 11-step sequence of the control and data lines needed to load a 16-bit value into a selected PIT counter's storage register. The process is almost identical to program mode except the value of A_1A_0 specifies which of the three counters is being loaded. The steps are:

(a) This is the normal, inactive state (which is the same for all sequences and is the condition of the control and data lines before and after all operations).

(b) CS* goes active and A_1A_0 contains the binary value of the counter to be loaded (00 for CTR_0, 01 for CTR_1 and 10 for CTR_2). (Recall that A_1A_0 equals 11 for the program mode operation.)

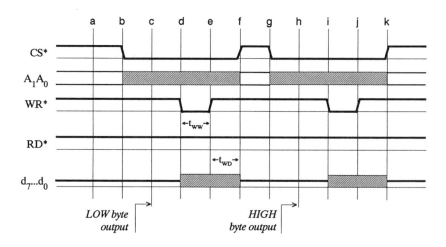

Figure 6.11. Sequence to load a value into a selected counter's storage register.

(c) Host software outputs the low byte of the load value. Because only 8 bits at a time can be transferred, the 16-bit value is divided into low and high bytes and value = 256 * high + low.

(d) WR* goes active and the low byte is put on the data lines.

(e) WR* returns inactive.

(f) The normal, inactive state is restored.

(g) CS* goes active again, beginning the second half of the operation. Again, A_1A_0 specifies the counter (which must be the same as in the first half).

(h) Host software outputs the high byte of the load value.

(i) WR* goes active and the high byte is put on the data lines.

(j) WR* returns inactive.

(k) The normal, inactive state is restored.

Critical times are the same as in the program mode sequence.

6.4.4 Sequence to Read a Selected Count Register

The timing diagram of Figure 6.12 shows the 11-step sequence of the control and data lines needed to read the 16-bit count value from a selected PIT counter. The steps are:

(a) This is the normal, inactive state of the control lines.

(b) CS* goes active and A_1A_0 has the binary value of the particular counter to be read.

(c) RD* goes active. The PIT responds to this action by putting on its data connections the binary value of the low byte of the selected counter's count register.

(d) Software inputs the low byte.

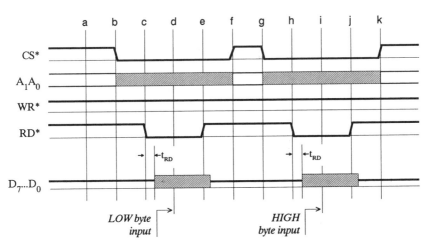

Figure 6.12. Sequence to read a selected count register.

(e) RD* returns inactive and the PIT removes the low byte from the data
     lines (they return to a Hi-Z state).
(f) The normal, inactive state is restored.
(g) CS* goes active again, beginning the second half of the read cycle. A_1A_0
     must have the same value as during the first half.
(h) RD* goes active for the second time. The high byte of the selected
     counter's count register is put on the data lines. PIT internal circuitry
     keeps track of read cycles and knows the most recent read got the low
     byte, and hence it's time for the high byte.
(i) Software inputs the high byte. The count value equals 256 * high + low.
(j) RD* returns inactive and the PIT responds by taking the data lines to a
     Hi-Z state.
(k) The normal, inactive state is restored.

The critical time, shown in Figure 6.12, is that data is not available for
input until at least t_{RD} = 300 ns after RD* goes active. (Consult Intel's specifications for further information.)

As a rule, the counting process should be stopped before reading the
count value. This is done by making the gate low or by stopping clock transitions. It's possible, however, to read the count value "on the fly"; that is to
latch the count value into a special register, and, while it is read, the regular
count register continues decrementing on high-low clock transitions. The
method is described in Intel's specifications.

6.5 Parallel Ports Interface

How can an 8253 be interfaced to the parallel ports so software can program,
load and read the three counters?

Just as memory IC's, PIT's require a bi-directional data bus $d_7...d_0$.
Therefore, it's again necessary to combine the data-out and data-in ports but
this time in a way consistent with PIT operations. Recall from Section 6.4
that when WR* is active, a value must go from the data-out port to the 8253
over $d_7...d_0$. Further, the value must remain until after WR* returns inactive.
Recall also that all PIT operations end with CS* returning inactive.

Figure 6.13 shows how to set up the bus. The 74373's Enable is wired
high which makes the latch value always equal data-out. An active WR*
clears the flip-flop and causes the 74373 to put data-out on $d_7...d_0$. The only
way to set the flip-flop and take data-out off the bus is for CS* to return
inactive.

The final issue is actually connecting the PIT to the parallel ports. Figure 6.14 gives a circuit in which:

$$WR^* = CO_1$$
$$RD^* = CO_2$$

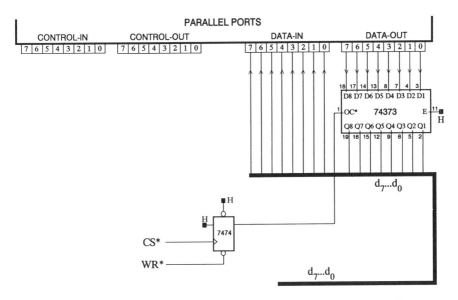

Figure 6.13. Circuit to produce a bi-directional data bus $d_7...d_0$. Data-out is put on the bus when WR* goes active and taken off when CS* returns inactive.

$$CS^* = CO_3$$
$$A_0 = CO_6$$
$$A_1 = CO_7$$

As was the case with the ADC and memory, it appears arbitrary which control-out bit supplies which control. However, the choices both earlier and here are consistent with the simultaneous control of PITs, memories and ADC's described in Chapter 7.

Finally, when CS* and RD* are active, the PIT puts a value on $d_7...d_0$ which is connected directly to $DI_7...DI_0$. Of course, RD* and WR* must never be active at the same time because both the PIT and 74373 would put values on the data bus.

6.6 Program, Load and Read Software

Once the circuit is set up, the next issue is how software programs, loads and reads the PIT. Specifically, what sequences of outputs to the control port in conjunction with outputs to the data port produce the program and load timing patterns given in Figures 6.10 and 6.11? And what sequence in conjunction

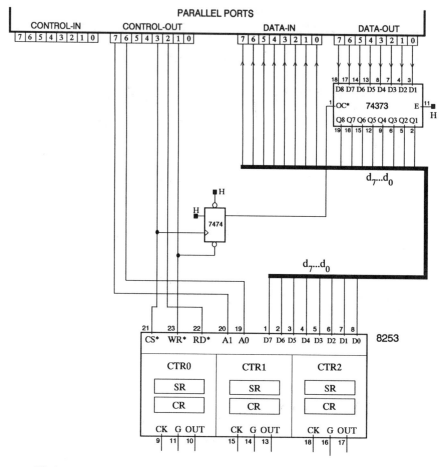

Figure 6.14. Circuit for interfacing and operating an 8253 through the parallel ports.

with inputs from the data port produce the read timing pattern given in Figure 6.12?

The outputs are determined by focusing on the relationship between the control-out bits and what they are connected to in the circuit.

$$CO_7 \quad CO_6 \quad CO_5 \quad CO_4 \quad CO_3 \quad CO_2 \quad CO_1 \quad CO_0$$
$$A_1 \quad A_0 \quad X \quad X \quad CS^* \quad RD^* \quad WR^* \quad X$$

The three X's are unused connections. Following an earlier convention, 1 is output for bits which don't matter.

6.6.1 Program PIT Mode

Figure 6.10 shows the timing sequence required to program the mode of an 8253 PIT counter. The following outputs to the control and data ports accomplish the sequence. (Sometimes the value of a control line doesn't matter, for example, $A_1 A_0$ at point a in the sequence. In those cases, 1 is output.)

point	CO_7 A_1	CO_6 A_0	CO_5 X	CO_4 X	CO_3 CS*	CO_2 RD*	CO_1 WR*	CO_0 X	hex value	constant name
a	1	1	1	1	1	1	1	1	FF	NORM
b	1	1	1	1	0	1	1	1	F7	SELPP
c	\multicolumn{10}{Output to data-port the Control Word (CW); this makes $DO_7...DO_0 = CW$}
d	1	1	1	1	0	1	0	1	F5	PROGP
e	1	1	1	1	1	1	1	1	FF	NORM

The following constants are defined:

NORM = &HFF	NORMal, inactive state
SELPP = &HF7	SELect PIT for Programming the mode (CS* active; $A_1 A_0 = 1\ 1$)
PROGP = &HF5	PROGram PIT mode (CS* and WR* active and $A_1 A_0 = 1\ 1$)
CP	Control Port number (value discussed in Chapter 3)
DP	Data Port number
CW	Control Word determines which PIT counter is programmed, what the mode is, etc. How to formulate values for CW is in Section 6.4.2.

Using the constants above, the following QuickBASIC subroutine programs whatever counter in whatever mode is specified by the parameter CW.

```
SUB PROGPIT(CW)
    OUT CP,NORM
    OUT CP,SELPP
    OUT DP,CW
    OUT CP,PROGP
    OUT CP,SELPP
    OUT CP,NORM
END SUB
```

As stated in Section 6.4.2, WR* (PROGP) must be active at least 400 ns. For fast computers, it may be necessary to output PROGP twice or even three times. Refer to Section 3.5.3 for the time between outputs.

6.6.2 Load PIT Counters

Load sequences are needed for each of the three counters (0, 1 and 2). Figure 6.11 shows the required timing pattern. The following outputs to the control and data ports accomplish the sequence for CTR_0.

point	CO$_7$ A$_1$	CO$_6$ A$_0$	CO$_5$ X	CO$_4$ X	CO$_3$ CS*	CO$_2$ RD*	CO$_1$ WR*	CO$_0$ X	hex value	constant name
a	1	1	1	1	1	1	1	1	FF	NORM
b	0	0	1	1	0	1	1	1	37	SELPC0
c	Output to data-port the low byte to be loaded into the storage register; this makes DO$_7$...DO$_0$ = LOW									
d	0	0	1	1	0	1	0	1	35	LOADPC0
e	0	0	1	1	0	1	1	1	37	SELPC0
f	1	1	1	1	1	1	1	1	FF	NORM
g	0	0	1	1	0	1	1	1	37	SELPC0
h	Output to data-port the high byte to be loaded into the storage register; this makes DO$_7$...DO$_0$ = HIGH									
i	0	0	1	1	0	1	0	1	35	LOADPC0
j	0	0	1	1	0	1	1	1	37	SELPC0
k	1	1	1	1	1	1	1	1	FF	NORM

The following additional integer constants are defined:

SELPC0	= &H37	**SEL**ect PIT Counter **0** (CS* active, $A_1A_0 = 0\ 0$)
LOADPC0	= &H35	**LOAD** PIT Counter **0** (CS* and WR* active, $A_1A_0 = 0\ 0$)
HIGH		**HIGH** 8 bits of the value to be loaded into the counter's storage register.
LOW		**LOW** 8 bits of the value to be loaded into the counter's storage register.
		HIGH = INT(VALUE/256)
		LOW = VALUE − HIGH*256

The following subroutine, using these constants, loads CTR_0 with value = 256 * HIGH + LOW.

```
SUB LOADPIT0(LOW,HIGH)
    OUT CP,NORM
    OUT CP,SELPC0
    OUT DP,LOW
    OUT CP,LOADPC0
    OUT CP,SELPC0
    OUT CP,NORM
    OUT CP,SELPC0
    OUT DP,HIGH
```

```
        OUT CP,LOADPC0
        OUT CP,SELPC0
        OUT CP,NORM
    END SUB
```

Load sequences for CTR_1 and CTR_2 are identical to load CTR_0 except the constants SELPC1, LOADPC1 and SELPC2, LOADPC2 replace SELPC0, LOADPC0. The new constants are the same as the ones they replace except in SELPC1, LOADPC1, $A_1A_0 = 0$ 1; and in SELPC2, LOADPC2, $A_1A_0 = 1$ 0. The values are:

	CO_7	CO_6	CO_5	CO_4	CO_3	CO_2	CO_1	CO_0	
	A_1	A_0	X	X	CS*	RD*	WR*	X	
SELPC1	0	1	1	1	0	1	1	1	&H77
LOADPC1	0	1	1	1	0	1	0	1	&H75
SELPC2	1	0	1	1	0	1	1	1	&HB7
LOADPC2	1	0	1	1	0	1	0	1	&HB5

The following two subroutines use these constants to load CTR_1 and CTR_2 with value = HIGH * 256 + LOW.

```
SUB LOADPIT1(LOW,HIGH)        SUB LOADPIT2(LOW,HIGH)
    OUT CP,NORM                   OUT CP,NORM
    OUT CP,SELPC1                 OUT CP,SELPC2
    OUT DP,LOW                    OUT DP,LOW
    OUT CP,LOADPC1               OUT CP,LOADPC2
    OUT CP,SELPC1                 OUT CP,SELPC2
    OUT CP,NORM                   OUT CP,NORM
    OUT CP,SELPC1                 OUT CP,SELPC2
    OUT DP,HIGH                   OUT DP,HIGH
    OUT CP,LOADPC1               OUT CP,LOADPC2
    OUT CP,SELPC1                 OUT CP,SELPC2
    OUT CP,NORM                   OUT CP,NORM
END SUB                       END SUB
```

In all three load subroutines, extend WR*, if necessary, by outputting second or third LOADPCi statements.

6.6.3 Read PIT Counters

Read sequences are needed for each of the three counters (0, 1 and 2). Figure 6.12 gives the timing pattern. The following outputs to the control and data ports accomplish the sequence for CTR_0:

point	CO_7 A$_1$	CO_6 A$_0$	CO_5 X	CO_4 X	CO_3 CS*	CO_2 RD*	CO_1 WR*	CO_0 X	hex value	constant name
a	1	1	1	1	1	1	1	1	FF	NORM
b	0	0	1	1	0	1	1	1	37	SELPC0
c	0	0	1	1	0	0	1	1	33	READPC0
d	In step c above, PIT CTR_0 puts the low byte of its current count register on $d_7...d_0$. In this step, that value is input from the data port and assigned to the variable LOW.									
e	0	0	1	1	0	1	1	1	37	SELPC0
f	1	1	1	1	1	1	1	1	FF	NORM
g	0	0	1	1	0	1	1	1	37	SELPC0
h	0	0	1	1	0	0	1	1	33	READPC0
i	In step h above, PIT CTR_0 puts the high byte of its current count register on $d_7...d_0$. In this step, that value is input from the data port and assigned to the variable HIGH.									
j	0	0	1	1	0	1	1	1	37	SELPC0
k	1	1	1	1	1	1	1	1	FF	NORM

One new constant is needed for the sequence:

READPC0   = &H33   **READ PIT** Counter **0** (CS* and RD* active, A_1A_0 = 0 0)

The following subroutine reads CTR_0 returning the count register value = HIGH * 256 + LOW:

```
SUB READPIT0(LOW,HIGH)
    OUT CP,NORM
    OUT CP,SELPC0
    OUT CP,READPC0
    LOW = INP(DP)
    OUT CP,SELPC0
    OUT CP,NORM
    OUT CP,SELPC0
    OUT CP,READPC0
    HIGH = INP(DP)
    OUT CP,SELPC0
    OUT CP,NORM
END SUB
```

Read sequences for CTR_1 and CTR_2 are identical to read CTR_0, except the constants SELPC1, READPC1 and SELPC2, READPC2 replace SELPC0, READPC0. The two new constants are identical to the ones they replace except in READPC1, A_1A_0 = 0 1; and in READPC2, A_1A_0 = 1 0. The actual values are:

	CO_7	CO_6	CO_5	CO_4	CO_3	CO_2	CO_1	CO_0	
	A_1	A_0	X	X	CS*	RD*	WR*	X	
READPC1	0	1	1	1	0	0	1	1	&H73
READPC2	1	0	1	1	0	0	1	1	&HB3

The following subroutines read CTR_1 and CTR_2 returning the contents of the count register:

```
SUB READPIT1(LOW,HIGH)        SUB READPIT2(LOW,HIGH)
    OUT CP,NORM                   OUT CP,NORM
    OUT CP,SELPC1                 OUT CP,SELPC2
    OUT CP,READPC1               OUT CP,READPC2
    LOW = INP(DP)                 LOW = INP(DP)
    OUT CP,SELPC1                 OUT CP,SELPC2
    OUT CP,NORM                   OUT CP,NORM
    OUT CP,SELPC1                 OUT CP,SELPC2
    OUT CP,READPC1               OUT CP,READPC2
    HIGH = INP(DP)                HIGH = INP(DP)
    OUT CP,SELPC1                 OUT CP,SELPC2
    OUT CP,NORM                   OUT CP,NORM
END SUB                       END SUB
```

As stated in Section 6.4.4, data cannot be input until at least 300 ns after RD* (READPCi) goes active. Depending on the times determined in Section 3.5.4, it may be necessary to output a second or third READPCi before inputting from the data port.

6.6.4 Sequence and Constant Summary

The seven sequences just described are all that's required to carry out PIT operations. Once set up as subroutines, they can be used over and over. The routines are called as follows:

```
CALL PROGPIT(CW)
CALL LOADPIT0(LOW,HIGH)
CALL LOADPIT1(LOW,HIGH)
CALL LOADPIT2(LOW,HIGH)
CALL READPIT0(LOW,HIGH)
CALL READPIT1(LOW,HIGH)
CALL READPIT2(LOW,HIGH)
```

The practice in this book is for every main program to declare all constants integer, assign them values, and share them with subroutines. The following statements accomplish this.

```
DEFINT A-Z
COMMON SHARED CP,DP
COMMON SHARED NORM, SELPP,PROGP
COMMON SHARED SELPC0,SELPC1,SELPC2
COMMON SHARED LOADPC0,LOADPC1,LOADPC2
COMMON SHARED READPC0,READPC1,READPC2
```

```
CP = &H300 : DP = &H301
NORM = &HFF
SELPP = &HF7 : PROGP = &HF5
SELPC0 = &H37 : SELPC1 = &H77 : SELPC2 = &HB7
LOADPC0 = &H35 : LOADPC1 = &H75 : LOADPC2 = &HB5
READPC0 = &H33 : READPC1 = &H73 : READPC2 = &HB3
```

Whenever real or other types of variables are needed, the DEFINT declaration is overridden by appending a character to variable names. In Quick-
BASIC, "!" designates single precision reals.

6.7 Troubleshooting Hardware and Software

Construct on a breadboard the PIT circuit in Figure 6.14. Refer to pin assignments in Appendix C and pin numbers in the figure. Carefully wire and
then check with a logic probe all +5's and grounds (power supply connections are not shown in the figure). Look over the rest of the wiring.

Enter the seven subroutines in Sections 6.6.1-6.6.3. Enter the constant
assignments and COMMON SHARED statements in Section 6.6.4. Save and
print the program shell. Then, proof and correct being especially thorough
with constant names and values.

It's essential to know the parallel ports work before beginning PIT tests.
Use software in Chapter 3 to check all 32 I/O bits.

*Recall that an infinite QuickBASIC loop is exited by typing CONTROL-BREAK.
Also, a program is aborted by typing CONTROL-C in response to an input.*

6.7.1 Test the Program PIT Subroutine

Add the following statements after COMMON SHARED and before the subroutines.

```
REPT:  CALL PROGPIT(0)            'CW = 0 is arbitrary
       GOTO REPT
       END
```

Save and then execute the program. Use an oscilloscope for the following
tests:
(1) Connect CS* (CO_3) to channel 1. It should activate once each loop. If
     not, check the wiring and constant names and values.
(2) Connect WR* (CO_1) to channel 2 (with CS* triggering the scope from
     channel 1). WR* should activate after and deactivate before CS* as
     shown in Figure 6.10. If not, check the wiring and then the software.
     Also, check that WR* is active at least 400 ns. If not, add OUT CP, PROGP
     statements.

(3)   Connect RD* (CO_2) to channel 2. It should always be high. If not, check the wiring and constants.

(4)   Next, connect A_0 (CO_6) and then A_1 (CO_7) to channel 2. Both should be high while CS* is active. If not, check the wiring and constants.

As stated in earlier troubleshooting sections, if a signal is wrong, disconnect it from whatever gates it is an input to and check again. If the signal then works, there is a wiring error or defective IC.

After the five controls work as expected, with CS* still on channel 1, look at the flip-flop output (OC*) on channel 2. It should go low when WR* goes active (after CS*) and return high when CS* goes high. If not, put a .01 μf noise capacitor between WR* (CLEAR*) and ground and check again. The 74373 must be properly controlled.

Finally, trigger the scope on OC* connected to channel 1 and look successively at the eight values of $d_7...d_0$ on channel 2. They should all be low. A good idea is to check values directly on PIT data pins. (If necessary, the data lines could be checked for wiring errors by redoing the program so the subroutine is called with other control word values.)

6.7.2 Test Load Pit Routines

Change the program to CALL LOADPIT0(0,0) rather than CALL PROGPIT(0). Check the five control lines as outlined above. This time CS* should go active twice each loop as shown in Figure 6.11. WR* should activate while CS* is active. A_1 and A_0 should both be low while CS* is active. (It isn't necessary to check the data lines or the flip-flop output again.) Be sure WR* is at least 400 ns long. Add OUT CP,LOADPC0 statements as needed.

Repeat the procedure with CALL LOADPIT1(0,0) and with CALL LOADPIT2(0,0). The only differences from LOADPIT0 are the values of A_1 and A_0. Fix the length of WR* as necessary.

Tedious testing is necessary in troubleshooting and always saves time.

6.7.3 Test Read PIT Subroutines

Change the program to CALL READPIT0(X,Y) rather than CALL LOAD-PIT2(0,0). Check the five control lines as outlined above. This time RD* should activate while CS* is active as shown in Figure 6.12. A_1 and A_0 should both be low. WR* should stay high. Check that RD* is active at least 600 ns. If not, add one or two OUT CP,READPC0 statements before the data inputs.

Carefully check that the flip-flop output (OC*) remains inactive during the loops. If not, add a .01 μf capacitor between WR* (CLEAR*) and ground.

Trigger the scope with RD* on channel 1. Successively look at the PIT's data lines on channel 2. The bit values are arbitrary but should be high or low (as opposed to Hi-Z).

Repeat the procedure with CALL READPIT1(X,Y) and with CALL READPIT2(X,Y). The only differences from READPIT0 are the values of A_1 and A_0. Fix the length of RD* as necessary.

6.7.4 Test PIT Operations

The last test is to actually program and load a PIT counter and observe its operation.

Connect a 1.0000 MHz digital oscillator to the clock input of CTR_0. Wire the gate to an initially low breadboard system logic switch. Connect the oscillator to scope channel 1 and CTR_0's output to channel 2.

The first exercise is to program mode 3 (rate generator) and load 10. This should make the output a 100.00 KHz symmetric clock. The mode control word from Section 6.4.2 is 00 11 011 0 = 36 hex. LOW is 10 and HIGH is 0.

Add the following statements to the program shell saved earlier:

```
CLS
PRINT "Program and Test PIT CTR0"
CALL PROGPIT(&H36)
CALL LOADPIT0(10,0)
INPUT "Programming and loading complete ";A$
END
```

Run the program. CTR_0's output, after the logic switch (connected to gate) is taken high, should be a clock with exactly one cycle for every 10 of the 1.0000 MHz oscillator. If the output is wrong, repeat the tests in Sections 6.7.1 and 6.7.2.

When everything works as expected, load different values and program other modes. This is an opportunity to explore PIT operations.

6.8 Example Application - A Scaler

The following application shows how to integrate several PIT counters to carry out timing and measurement operations. It also illustrates additional software techniques.

The application is a manually controlled "scaler" which determines the number of events received during a 1.0000 second interval. The inputs to the scaler are an active high pulser (which initiates a counting interval) and an events signal (with each event producing a high-low transition).

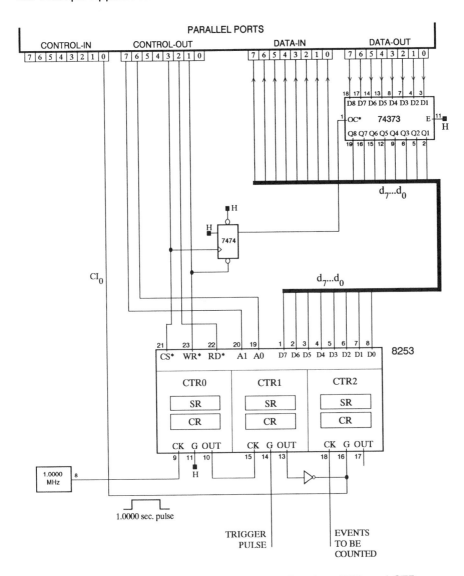

Figure 6.15. Circuit for a manually triggered scaler. CTR_0 and CTR_1 produce a 1.0000 second pulse which after inversion supplies the gate to CTR_2. CTR_2 counts events.

6.8.1 Hardware

The hardware must upon trigger produce a 1.0000 second long active-high pulse during which events are counted. The complete circuit is given in Figure 6.15. The PIT is wired to the parallel ports as previously shown in Figure 6.14. The only new circuitry involves the three counters' clock and gate inputs and outputs, a 1.0000 MHz master clock, and the control-in port's CI_0.

Three PIT counters are required. CTR_1, programmed for mode 1, produces a 1.0000 second active low output pulse. The loaded value$_1$ times the input clock$_1$'s period must equal 1.0000 second. The pulse (output$_1$) is triggered by a manually controlled active-high pulser connected to gate$_1$.

CTR_2, programmed in mode 2, counts events. The pulse from CTR_1 is inverted to active high and applied to gate$_2$. The events signal is connected to clock$_2$. Value$_2$ is 7FFF hex. While gate$_2$ is high (1.0000 second), each event (after the first low-high-low of clock$_2$) decrements count register$_2$ by 1.

The status of gate$_2$ is monitored at control-in bit 0 (CI_0). Software can then determine when the 1.0000 second interval begins and subsequently ends, and therefore when to read the decremented count register.

The master clock is a 1.0000 MHz crystal oscillator which produces a symmetric square wave. The period is 1.0000×10^{-6} seconds. One million periods are needed to make 1.0000 second. Since the largest possible PIT value is 65535, it's necessary for CTR_0 to divide the 1.0000 MHz down. Of many possibilities, suppose CTR_0 is programmed for mode 3 (square-wave generation) and loaded with 1000. Output$_0$'s period is $1000 * (1.0000*10^{-6}) = 1.0000 \times 10^{-3}$ seconds = 1 ms (f = 1000 Hz). Then 1000 output$_0$ cycles is a 1.0000 second pulse.

The following table shows the plan.

PIT counter	purpose	clock input	gate input	output	loaded value	mode
0	divide 1.0000 MHz down to 1000.0 Hz	1.0000 MHz sq wave	wired high	1000.0 Hz sq wave	1000	3
1	produce 1.0000 second pulse	1000.0 Hz from CTR_0	trigger pulser	1.0000 sec. pulse	1000	1
2	count events	event signal	inverted pulse from CTR_1	not used	7FFF hex	2

6.8.2 Software

Scaler software has three parts: initialization, polling and reading data. Each is described below after which the complete program is given.

Initialization

First, program the mode of all three counters. This requires three calls of the PROGPIT(CW) subroutine given earlier. But first, the control word (CW) for each counter must be determined. How to do so was covered earlier and the table below gives CW values.

Next, load values into the three counters. This requires calling the three LOADPITi(LOW,HIGH) subroutines described earlier. But first, the load value's LOW and HIGH bytes are determined. Recall that value = (HIGH * 256) + LOW which means:

```
HIGH = INT(VALUE/256)        and
LOW = VALUE - HIGH*256
```

PIT counter	mode	program mode CW	mode constant name	loaded value	LOW	LOW constant name	HIGH	HIGH constant name
0	3	00110110	CW0=&H36	1000	232	LOW0	3	HIGH0
1	1	01110010	CW1=&H72	1000	232	LOW1	3	HIGH1
2	2	10110100	CW2=&HB4	&H7FFF	&HFF	LOW2	&H7F	HIGH2

Note that 3*256 + 232 = 1000 and INT(1000/256) = 3.

There is a good reason &H7FFF is the value loaded into CTR_2 (which counts events). In most BASIC's (including Microsoft QuickBASIC), single precision integer numbers are 16 bits. The largest positive value is 32767 = 7FFF hex. Larger hex numbers translate to negative integer values. Therefore, 7FFF is a reasonable load value to decrement from. If the value is larger and there are only a small number of events in an interval, then (value - read count value) + 1 may still be negative, something which should be avoided.

The complete program including the initialization described here is given later.

Polling

After initializing the PIT's three counters, software outputs "Initialization Complete. Activate Pulser when Ready." Next, the program waits until data is ready. Figure 6.15 shows that the signal at $gate_2$ (which is high only during the counting interval) also supplies CI_0. Software monitors the bit to determine when the interval starts and stops. Two loops are required. The first Waits While CI_0 is Low (waits until the scaler is manually triggered). The second Waits While CI_0 is High (CTR_2 is counting). Data is ready when the second loop ends.

The new constant MASK = &H01 is used in the loops.

If the counting interval is sufficiently short that no monitor inputs might occur while CI_0 is high, then additional circuitry is required as discussed in Section 4.4.

Read Count Register Value

When the wait loops end, it's time to read the number of events counted during the 1.0000 second interval. This is done with the read sequence for CTR_2 given earlier. As stated, the count register = HIGH * 256 + LOW and the number of events = &H7FFF - count register + 1.

Repeat

Finally, after an interval, software supports a repeat option. If selected, execution branches back to the start of polling. It's not necessary to reinitialize because values are reloaded from the storage to the count registers when the pulser is again activated.

Complete Scaler Program

```
'        Program SCALER
'        Standard constants and declarations
         DEFINT A-Z
         COMMON SHARED CP,DP
         COMMON SHARED NORM, SELPP,PROGP
         COMMON SHARED SELPC0,SELPC1,SELPC2
         COMMON SHARED LOADPC0,LOADPC1,LOADPC2
         COMMON SHARED READPC0,READPC1,READPC2

         CP = &H300 : DP = &H301
         NORM = &HFF
         SELPP = &HF7 : PROGP = &HF5
         SELPC0 = &H37 : SELPC1 = &H77 : SELPC2 = &HB7
         LOADPC0 = &H35 : LOADPC1 = &H75 : LOADPC2 = &HB5
         READPC0 = &H33 : READPC1 = &H73 : READPC2 = &HB3
'        New constants for the SCALER program
         CW0 = &H36 : LOW0 = 232 : HIGH0 = 3
         CW1 = &H72 : LOW1 = 232 : HIGH1 = 3
         CW2 = &HB4 : LOW2 = &HFF : HIGH2 = &H7F
         MASK = &H01
'        Initialization
         CALL PROGPIT(CW0)
         CALL PROGPIT(CW1)
         CALL PROGPIT(CW2)

         CALL LOADPIT0(LOW0,HIGH0)
         CALL LOADPIT1(LOW1,HIGH1)
         CALL LOADPIT2(LOW2,HIGH2)
         PRINT "Initialization Complete"
'        Main Program Loop
MLOOP:   PRINT "Activate Pulser When Ready"
WWL:     IF (INP(CP) AND MASK) = 0 THEN GOTO WWL
         PRINT "Interval Starting"
```

```
WWH:    IF (INP(CP) AND MASK) = MASK THEN GOTO WWH
'       Data ready to be read and counts computed
        CALL READPIT2(LOW,HIGH)
        CR = HIGH*256 + LOW
        EVENTS = &H7FFF - CR + 1
        PRINT "Number of Events: "; EVENTS

'       Repeat option
        INPUT "Type <ENTER> to Repeat, E to END ";A$
        IF A$ = "E" OR A$ = "e" THEN GOTO FINI
        GOTO MLOOP
FINI:   END

        Text of the seven subroutines given in
        Sections 6.61, 6.62 and 6.63.
```

6.8.3 Comments and a Look Ahead

The manually triggered scaler demonstrates PIT and parallel port operations in a straightforward context. Further, the application provides a framework for considering several broader issues involved in the design of more powerful data acquisition circuits.

The scaler shows the advantages of software in the implementation of instruments. For example, suppose a different counting interval is desired. Not one wire need be changed. Just load different initial values into CTR_0 and/or CTR_1. In fact, with the following software, the interval is an input variable:

```
PRINT "Enter the Desired Counting Interval in MS"
INPUT "Range (1 to 10000) : ";NOMS
HIGH = INT(NOMS/256) : LOW = NOMS - HIGH*256
```

After high and low are sent to the load CTR_1 routine, the counting interval is NOMS ms long. (Recall that the clock input to CTR_1 is 1.0000 ms.)

Suppose one wants to acquire data from a large number of intervals. This could almost totally be done in software by surrounding the polling and read operations with a loop which assigns the data value to an array element each time around. However, a limitation is that each counting interval is manually triggered. This could be overcome if the hardware supported a way for software to trigger an interval. Implementation of a software start/stop is described in Chapter 7, and a software triggered scaler is presented in Chapter 9.

Suppose the goal is to determine the number of events received during a succession of intervals in real time. A multiscaler can be built by adding a fourth PIT counter (which means adding another 8253 to the system). The idea is for CTR_A (the current CTR_2) to count during the first interval, CTR_B (in the added PIT) during the second, CTR_A again during the third, CTR_B

during the fourth and so on. After the first interval, while CTR_B is counting, CTR_A is read; then when CTR_A is counting again, CTR_B is read; and so on. There is no dead time between the end of one interval and the start of the next. If real time begins with the first interval, then, at the end of the nth interval, real time equals n multiplied by the interval length. How to add an additional PIT (actually up to seven 8253's) is described in Chapter 7. Complete circuitry and software for a single-channel real time software or hardware triggered multiscaler is given in Chapter 9.

6.9 Troubleshooting the Scaler

Construct the scaler circuit in Figure 6.15, or modify the circuit in Section 6.7 to be the scaler. Use pin assignments in the figure and in Appendix C. Use a breadboard system pulser for the trigger and the system clock for the events signal. Set the frequency on 100 or 1000 Hz. Check +5 and ground connections with a logic probe. Inspect the wiring.

Enter the scaler program. Don't forget the seven subroutines. Save, print, proof and correct as necessary. Run the program. If the system doesn't work, do the troubleshooting sequence in Section 6.7.

After successfully completing the tests, the scaler should work. If not, the most likely problem is the new constants. Check and correct.

Next, run the program with a logic probe on CTR_2's gate input (CI_0). Activate the trigger and observe the 1.0000 second pulse. If it does not occur, check the scaler wiring, CTR_0's output and CTR_1's gate. If the logic probe shows the pulse but the program hangs up, check the polling software.

Finally, if the system seems to work but displays clearly wrong values, temporarily put INPUT A\$ statements in the READPIT2 subroutine after each INP(DP). Run the program. When execution stops the first time, use the logic probe to determine the binary value of LOW. When execution stops again, determine HIGH. Correct values are roughly predictable from the loaded numbers and the breadboard clock's frequency.

Once the scaler works, variations such as those in several exercises can be explored.

6.10 A Potential Fatal Fault

The scaler's hardware and software described in Section 6.8 fails if there is not at least one low-high-low on the events line during a counting interval. As previously explained, 7FFF hex is initially loaded into CTR_2's storage register and is not transferred to the count register until after its $gate_2$ (1.0000 second pulse) goes high and its clock input (events signal) goes low-high-low. That's why the first event is not counted and 1 is added to read values. If there are no events, the count register is never loaded and the read operation yields indeterminate results. Therefore, if no events is a possibility, measures must be taken.

If the electrical representation of events is restricted to "short" low-high-low pulses, then "no" events could be dealt with by simply ORing a low-high-low pulse with the events line and feeding the results to $clock_2$. This works because the OR gate passes the pulse as long as the events line is low. (With pulse transferring the load value from the storage to the count register, it's no longer necessary to add 1 to the number of high-low transitions counted.)

The remaining question is how to originate the pulse. A hardware solution is to add an 8253 and generate the pulse with the one-shot mode. A software solution is also possible. Suppose $CO_7 = A_1$ is inverted and ORed with the events signal. The sequence of outputs below produces the needed pulse. (The reason for the inversion is simply to maintain high as the inactive state.)

        OUT   1111 1111 CO_7 after inversion   0
        OUT   0111 1111   1
        OUT   1111 1111   0

The sequence must be located after the first polling loop which is reached when $gate_2$ has just gone high. It's safe to use A_1 to make the pulse because $gate_2$ is low during programming, loading and reading operations (when A_1 is used for other purposes).

A requirement of the pulse solution is that the events line be low when there are no events. If the line is high, simply ORing the pulse won't work because the OR output would be high before, during and after pulse.

A possible solution in this case uses two pulses. P_1 is active low and ANDed with the events line. When P_1 is low, the line is masked out. The AND output is ORed with a second active high pulse P_2 and the result sent to $clock_2$. P_2 must begin after and end before P_1. The whole scheme is shown in

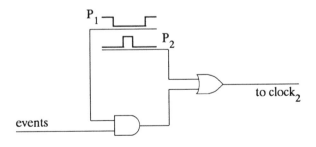

Figure 6.16. The addition of a low-high-low pulse P_2 to the events line solves the problem of no events. P_1 suppresses the line so P_2 gets through. P_1 and P_2 may be generated by software or hardware.

Figure 6.16. The two pulses may be generated by two or three PIT counters or by software. In the latter case, P_1 might be A_1 and P_2 might be A_0 inverted.

For the scaler to work in all event line situations, the two pulse or some other circuit addition is required. But if an application always produces events, no solution is needed. And if the events line is low when there are no events, then a simpler solution is sufficient.

Exercises

1. Use parallel ports (which meet specifications in Chapter 3) and a breadboard system to carry out the testing and troubleshooting sequence in Section 6.7.

2. After doing the exercise in Section 6.7.4, write software which programs PIT CTR_0 for mode 1 and loads 10. Connect the breadboard system clock, set on 1000 Hz, to the gate. What should the output be? What is it?

3. Write a short program which tests the subroutine READPIT0. Devise a way to use the breadboard system clock and a pulser to verify the hardware and software. Don't forget to initially program mode 2 and load an initial value.

4. Use Intel's specifications and the troubleshooting circuit and software to explore PIT modes 0, 4 and 5.

5. Build, test and troubleshoot the scaler described in Sections 6.8 and 6.9.

6. Write a version of the scaler program which inputs the counting interval length and after data acquisition displays the frequency of the signal at $clock_2$.

7. Design a software trigger for the scaler (both the extra hardware, if any, and software).

8. How might the scaler fail if the counting interval is less than 1 ms? Design a solution for this potential problem.

9. Devise a frequency meter in which software repeatedly determines and displays the frequency of the signal applied to CTR_2's clock input. The simplest approach is to program the counting interval length in anticipation of the frequencies to be determined with the goal of getting as

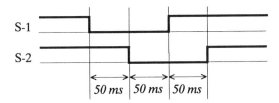

Figure 6.17. Figure for Exercise 12.

many counts as possible without going over 32K, the largest positive 16-bit integer value. Another approach is to start with a standard interval and then reprogram according to what's found. For both approaches, software might reprogram the interval as the frequency changes and thereby maximize the range which can be determined..

10. Design a circuit which uses four PIT counters (two 8253 IC's) to produce a real-time multiscaler along the lines discussed in Section 6.7.3.

11. Design a circuit and write software which uses several PIT counters to produce upon trigger a signal which consists of exactly 1000 cycles of a 1.000 ms (period) clock. Start with the 1.0000 MHz digital oscillator.

12. Design a circuit and write software which uses some number of PIT counters to produce the two timing signals specified in Figure 6.17. Start with a 1.0000 MHz digital oscillator.

13. Add a PIT to the ADC circuit in Figure 4.4 so the rate of voltage conversion can be programmed by software. Two of the three counters might generate a rate clock and the third produce the convert pulse. Can any of the control lines serve both the ADC and PIT? Recall that a PIT does not respond to RD*, WR*, A_1 and A_0 unless CS* is active.

14. How would you set up and program an 8253 to generate P_1 and P_2 discussed in Section 6.10?

7 Control/Data Interface

Chapters 4-6 are tutorials on three specific devices: the AD573 Analog-to-Digital Converter, the 6264 static RAM and the Intel 8253 Programmable Interval Timer. The devices by themselves are insufficient for reasonable data acquisition systems. For instance, it's desirable to use several 8253's for counting and timing measurements (as discussed in Section 6.8.3), and it's advantageous to use a PIT to control RAM storage of data (as discussed in Section 5.6.7). This chapter describes a framework which supports multiple devices.

The **Parallel Data Collector** (hereafter called the PDC) is a modular hardware and software system which facilitates the design, development, testing and use of a variety of data acquisition systems. The parallel ports described in Chapter 3 comprise the first module and is the only part of the system including software which depends on the particular host computer. A second module, the Control/Data Interface (CDI), produces from the ports a control and data structure sufficiently elaborate to support multiple PIT's, memories and/or ADC's (as well as other category devices). A third module contains two ADC IC's. The plan is to use the parallel ports, CDI and ADC as service modules over and over with a variety of applications. Figure 7.1 is a block diagram of the system.

This chapter describes the CDI including specifications, design, software, construction and testing. Chapter 8 presents the ADC. Both modules are built on individual wire wrap cards, and complete drawings are provided. Specific card rack, power supply and interconnecting cables are recommended.

After the framework is established, a general purpose measurement system is presented in Chapter 9. And a "fast" voltage recorder is given in Chapter 10. The chapters show how to create specific measurement capabilities and provide a background for the design and implementation of other applications.

7.1 Overview

Chapters 4-6 show how to interface PIT, memory and ADC devices to the parallel ports. Bits from the control-out port ($CO_7...CO_0$) supply signals which

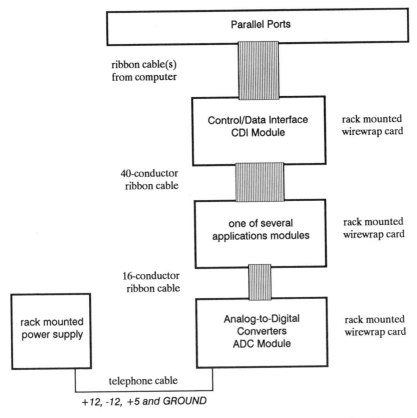

Figure 7.1. Block diagram of the Parallel Data Collector (PDC).

program, load and read the devices. Recall that a single 8253 requires five controls, the 6264 memory circuit also needs five and a pair of AD573's needs at least four. Obviously, the 8 control-out bits are insufficient for multiple devices. To solve this problem and others, the CDI:

(1) Generates A_1, A_0, RD*, WR* and seven independent CS* controls for 8253 PIT's and other devices.
(2) Sets up, from the data-out and data-in ports, a bi-directional data bus $d_7...d_0$ suitable for PIT, memory and other devices.
(3) Makes available four memory controls (the fifth in Chapter 5 proves unnecessary).
(4) Develops five independent read enables suitable for AD573's and other devices.
(5) Produces a software Start/Stop (S/S*).
(6) And sends 10 monitor lines to control-in.

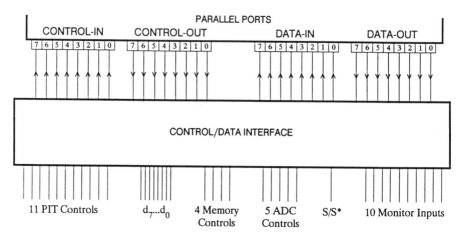

Figure 7.2. Block diagram of CDI operations.

In essence, the CDI converts the 32 bits of the parallel ports to the more efficient format summarized in Figure 7.2.

> *These particular capabilities were arrived at with the specific applications in Chapters 9 and 10 in mind. More powerful CDI's are certainly possible and may be necessary for conceivable applications. Exercises at the end of this chapter explore adding capabilities.*

The primary CDI circuit elements are two 74138 decoder chips. One generates PIT and memory controls and the other produces ADC controls and S/S*.

> *Recall from Section 2.8.1 that the 74138 has eight outputs. When its three gates are active, output Y_i is low where i is the binary value of the select inputs CBA. All other outputs are high. And all outputs are high if any gate is inactive.*

> *Controls are characterized as active and inactive rather than high and low. Recall that a signal name with an appended "*" is active-low. And no "*" usually means active-high.*

The 74138's (labeled A and B) are connected to control-out as shown in Figure 7.3. CO_3 is wired to both select A's, CO_4 to both select B's and CO_5 to both C's. CO_0 is connected to G_1 of 74138-A and to G_{2A}* and G_{2B}* of 74138-B. All other gates are wired active. When CO_0 is high, 74138-A responds to $CO_5 CO_4 CO_3$; and when CO_0 is low, 74138-B responds. Only one output from either chip can be low at a time.

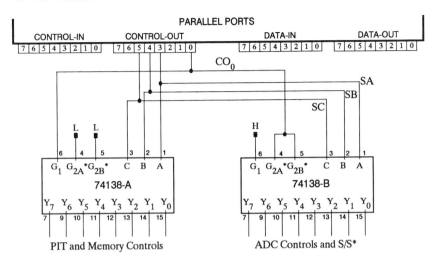

Figure 7.3. Two 74138's are wired to control-out so 74138-A works when CO_0 is high and 74138-B when CO_0 is low.

Sections 7.2 and 7.3 show how 74138-A produces PIT and memory controls. Sections 7.4 and 7.5 show how 74138-B generates ADC controls and S/S*. Section 7.6 presents the control-in's. Section 7.7 reviews the entire circuit. Section 7.8 gives complete construction details. And Section 7.9 covers testing and troubleshooting.

7.2 PIT Controls

7.2.1 Hardware

Section 6.4.1 describes the five controls needed to program, load and read 8253 PIT's. Recall that CS* (Chip Select) must be active for the chip to respond to RD* (ReaD), WR* (WRite), A_1 (Address 1) and A_0 (Address 0) commands. Therefore, multiple 8253's could all use the same RD*, WR*, A_1 and A_0 as long as each has a unique CS*. 74138-A generates seven independent Chip Selects (designated CS_0* to CS_6*) as shown in Figure 7.4. When CO_0 is high, only CS_i* is active where $i = CBA = CO_5CO_4CO_3$.

Y_7 (an eighth CS*) is not used for reasons discussed later.

Recall that PIT's have bi-directional data connections. The CDI combines the data-in and data-out ports to produce a bus, $d_7...d_0$, which fulfills specifications in Section 6.5. Circuitry is in Figure 7.5. A 74244 buffer (rather than a 74373) is used for reasons discussed in Section 7.7.

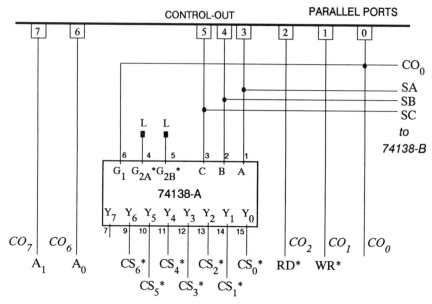

Figure 7.4. Circuitry which generates PIT controls.

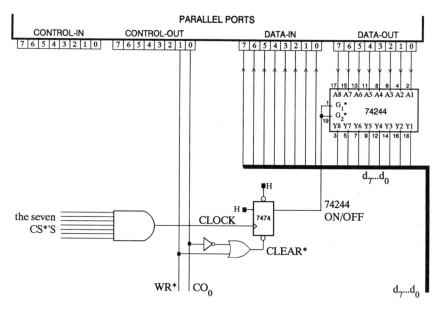

Figure 7.5. Circuitry which produces a bi-directional data bus. WR*
turns the bus on and one of the CS*'s turns it off.

Recall from Section 2.9.2 that when a 74244's gates G_1* and G_2* are active, the outputs equal the corresponding inputs. When the gates are inactive, the outputs are in a Hi-Z state.

The 74244 is turned on and off as follows.

When CO_0 is low (for read ADC operations), one input to the OR gate is high and the flip-flop's CLEAR* stays high no matter what WR* does. On the other hand, when CO_0 is high (for PIT operations), the OR input is low and CLEAR* is the same as WR*. In this case, WR*'s activation clears the flip-flop and turns the 74244 on (as required). WR*'s deactivation has no effect.

Only one CS* can be active at a time. The AND output (flip-flop CLOCK input) is high when no CS* is active. Activation of any CS* takes CLOCK low. Deactivation takes CLOCK back high and, since DATA is wired high, sets the flip-flop. This turns the 74244 off as required.

7.2.2 Software Overview

Circuitry in Figures 7.4 and 7.5 allows multiple PIT's to be interfaced to the parallel ports and individually programmed, loaded and read. Suppose seven 8253's (numbered 0 to 6) are all wired to WR*, RD*, A_0, A_1 and $d_7...d_0$. Suppose CS_0* goes to PIT_0, CS_1* to PIT_1, etc. The connections are summarized in the following table.

CO_7	CO_6	CO_5	CO_4	CO_3	CO_2	CO_1	CO_0	
A_1	A_0	C	B	A	RD*	WR*	1	CS_{CBA}* is active

The following examples illustrate the action of bytes sent to control-out. As always, X means high or low.

CO_7	CO_6	CO_5	CO_4	CO_3	CO_2	CO_1	CO_0	Action of output
A_1	A_0	C	B	A	RD*	WR*	1	
X	X	X	X	X	X	X	0	Has no effect on the PIT's
X	X	0	0	0	X	X	1	Activates CS_0* (CBA = 0)
X	X	1	0	0	X	X	1	Activates CS_4* (CBA = 4)
X	X	1	1	1	X	X	1	Has no effect (Y_7 is not used)
X	X	0	1	1	1	0	1	Activates CS_3* and WR* and turns the 74244 on
X	X	0	1	1	1	1	1	Keeps CS_3* active and the 74244 on
X	X	1	1	1	1	1	1	Activates no CS* but turns the 74244 off if the last two outputs precede this one

In Sections 6.6.1-6.6.4, constants and software were developed which program, load and read the three independent counters in a single PIT. The same thing is done here for up to seven PIT's. The first task is to determine the output which puts all controls in the **NORM**al, inactive state shown at points **a** in Figures 6.10, 6.11 and 6.12. No CS* is active if the 74138-A's select inputs are 7 or 111 binary. (This is one reason Y_7 is not used.) CO_0 is high for all PIT operations. RD* and WR* are normally inactive. A_1 and A_0 may be high or low. With the convention adopted earlier to output 1 for bits which don't matter, NORM is:

CO_7	CO_6	CO_5	CO_4	CO_3	CO_2	CO_1	CO_0	hex	constant
A_1	A_0	C	B	A	RD*	WR*	1	value	name
1	1	1	1	1	1	1	1	FF	NORM

NORM has the same value as in Chapter 6. Again, it's desirable for inactive to be the same as disconnected which in TTL logic is high.

7.2.3 Software to Program PIT's

The **PROG**ram **PIT** subroutine is reproduced from Section 6.6.1.

```
SUB PROGPIT(CW)
   OUT CP,NORM
   OUT CP,SELPP
   OUT DP,CW
   OUT CP,PROGP
   OUT CP,SELPP
   OUT CP,NORM
END SUB
```

When executed, the output sequence implements the timing pattern in Figure 6.10. CP and DP are the port numbers. CW is the mode Control Word. SELPP (**SEL**ect **PIT** for Programming) and PROGP (**PRO**gram **PIT**) are constants whose values are assigned in the calling program and declared COM-MON SHARED. SELPP establishes point **b** in Figure 6.10 and is the same as NORM except CS* is active (A_1 and A_0 are already high). PROGP establishes point **d** and is the same as SELPP except WR* is active.

In the current case, SELPP and PROGP must have different values for different PIT's. The problem is how to efficiently set up the constants. Suppose arrays are declared.

```
DIM SELPP(7)
DIM PROGP(7)
```

Next, suppose values are assigned so that SELPP(0) and PROGP(0) activate PIT_0 (CBA = 000), SELPP(1) and PROGP(1) activate PIT_1 (CBA = 001), etc. With the assignments made in and shared by the calling program, the following subroutine programs any of seven 8253's.

```
SUB PROGPIT(I,CW)
   OUT CP,NORM
   OUT CP,SELPP(I)
   OUT DP,CW
   OUT CP,PROGP(I)
   OUT CP,SELPP(I)
   OUT CP,NORM
END SUB
```

The new parameter I is the PIT number (0 to 6). The routine is called by specifying I and the mode control word. For example,

```
CALL PROGPIT(4,&HB6)
```

programs PIT number 4 with CW = B6 hex. During execution, the routine uses SELPP(4) and PROGP(4).

The next job is to determine values for the constants. SELPP(I) is the same as SELPP in Chapter 6 except I (=CBA) specifies which CS* is active. The values are:

CO_7	CO_6	CO_5	CO_4	CO_3	CO_2	CO_1	CO_0	hex	constant
A_1	A_0	C	B	A	RD*	WR*	1	value	name
1	1	0	0	0	1	1	1	C7	SELPP(0)
1	1	0	0	1	1	1	1	CF	SELPP(1)
1	1	0	1	0	1	1	1	D7	SELPP(2)
1	1	0	1	1	1	1	1	DF	SELPP(3)
1	1	1	0	0	1	1	1	E7	SELPP(4)
1	1	1	0	1	1	1	1	EF	SELPP(5)
1	1	1	1	0	1	1	1	F7	SELPP(6)

PROGP(I) is the same as SELPP(I) except WR* is active. The values are:

CO_7	CO_6	CO_5	CO_4	CO_3	CO_2	CO_1	CO_0	hex	constant
A_1	A_0	C	B	A	RD*	WR*	1	value	name
1	1	0	0	0	1	0	1	C5	PROGP(0)
1	1	0	0	1	1	0	1	CD	PROGP(1)
1	1	0	1	0	1	0	1	D5	PROGP(2)
1	1	0	1	1	1	0	1	DD	PROGP(3)
1	1	1	0	0	1	0	1	E5	PROGP(4)
1	1	1	0	1	1	0	1	ED	PROGP(5)
1	1	1	1	0	1	0	1	F5	PROGP(6)

These constants and values are summarized in Section 7.7.3.

7.2.4 Software to Load PIT's

The subroutine which **LOAD**s **PIT** counter **0** is reproduced from Section 6.6.2.

```
SUB LOADPIT0(LOW,HIGH)
  OUT CP,NORM
  OUT CP,SELPC0
  OUT DP,LOW
  OUT CP,LOADPC0
  OUT CP,SELPC0
  OUT CP,NORM
  OUT CP,SELPC0
  OUT DP,HIGH
  OUT CP,LOADPC0
  OUT CP,SELPC0
  OUT CP,NORM
END SUB
```

When executed, the sequence of outputs produces the timing pattern in Figure 6.11. LOW and HIGH are loaded into CTR_0 (the actual 16-bit value is HIGH*256 + LOW). The constants are SELPC0 (**SEL**ect **PIT** Counter **0**) and LOADPC0 (**LOAD** a byte in **PIT** Counter **0**). In Chapter 6, separate constants and subroutines load CTR_1 and CTR_2.

The issue now is to determine the best way to program up to three counters in up to seven 8253's. Suppose SELPC(I,J) and LOADPC(I,J) are declared two-dimensional arrays with the index I specifying the PIT and J the counter. I goes from 0 to 6 and J from 0 to 2.

```
DIM SELPC(7,3)
DIM LOADPC(7,3)
```

After assigning values to the constants, the following routine loads any counter in any PIT.

```
SUB LOADPIT(I,J,LOW,HIGH)
  OUT CP,NORM
  OUT CP,SELPC(I,J)
  OUT DP,LOW
  OUT CP,LOADPC(I,J)
  OUT CP,SELPC(I,J)
  OUT CP,NORM
  OUT CP,SELPC(I,J)
  OUT DP,HIGH
  OUT CP,LOADPC(I,J)
  OUT CP,SELPC(I,J)
  OUT CP,NORM
END SUB
```

The routine is called by specifying PIT and counter numbers and the LOW and HIGH bytes to load. For example,

```
CALL LOADPIT(5,1,10,0)
```

loads PIT 5, counter 1 with 10. During execution, the routine uses SELPC(5,1) and LOADPC(5,1).

The next job is to determine values for the constants. The objective of outputting SELPC(I,J) is to produce point **b** in Figure 6.11. A_1 and A_0 specify the counter in the 8253 whose CS* is active. RD* and WR* are inactive. The values are:

CO_7	CO_6	CO_5	CO_4	CO_3	CO_2	CO_1	CO_0	hex	constant
A_1	A_0	C	B	A	RD*	WR*	1	value	name
0	0	0	0	0	1	1	1	07	SELPC(0,0)
0	1	0	0	0	1	1	1	47	SELPC(0,1)
1	0	0	0	0	1	1	1	87	SELPC(0,2)
0	0	0	0	1	1	1	1	0F	SELPC(1,0)
0	1	0	0	1	1	1	1	4F	SELPC(1,1)
1	0	0	0	1	1	1	1	8F	SELPC(1,2)
0	0	0	1	0	1	1	1	17	SELPC(2,0)
0	1	0	1	0	1	1	1	57	SELPC(2,1)
1	0	0	1	0	1	1	1	97	SELPC(2,2)
0	0	0	1	1	1	1	1	1F	SELPC(3,0)
0	1	0	1	1	1	1	1	5F	SELPC(3,1)
1	0	0	1	1	1	1	1	9F	SELPC(3,2)
0	0	1	0	0	1	1	1	27	SELPC(4,0)
0	1	1	0	0	1	1	1	67	SELPC(4,1)
1	0	1	0	0	1	1	1	A7	SELPC(4,2)
0	0	1	0	1	1	1	1	2F	SELPC(5,0)
0	1	1	0	1	1	1	1	6F	SELPC(5,1)
1	0	1	0	1	1	1	1	AF	SELPC(5,2)
0	0	1	1	0	1	1	1	37	SELPC(6,0)
0	1	1	1	0	1	1	1	77	SELPC(6,1)
1	0	1	1	0	1	1	1	B7	SELPC(6,2)

The objective of outputting LOADPC(I,J) is to establish point **d** in Figure 6.11. The only difference from SELPC(I,J) is that WR* is active. The values are:

CO_7	CO_6	CO_5	CO_4	CO_3	CO_2	CO_1	CO_0	hex	constant
A_1	A_0	C	B	A	RD*	WR*	1	value	name
0	0	0	0	0	1	0	1	05	LOADPC(0,0)
0	1	0	0	0	1	0	1	45	LOADPC(0,1)
1	0	0	0	0	1	0	1	85	LOADPC(0,2)
0	0	0	0	1	1	0	1	0D	LOADPC(1,0)
0	1	0	0	1	1	0	1	4D	LOADPC(1,1)
1	0	0	0	1	1	0	1	8D	LOADPC(1,2)

0	0	0	1	0	1	0	1	15	LOADPC(2,0)
0	1	0	1	0	1	0	1	55	LOADPC(2,1)
1	0	0	1	0	1	0	1	95	LOADPC(2,2)
0	0	0	1	1	1	0	1	1D	LOADPC(3,0)
0	1	0	1	1	1	0	1	5D	LOADPC(3,1)
1	0	0	1	1	1	0	1	9D	LOADPC(3,2)
0	0	1	0	0	1	0	1	25	LOADPC(4,0)
0	1	1	0	0	1	0	1	65	LOADPC(4,1)
1	0	1	0	0	1	0	1	A5	LOADPC(4,2)
0	0	1	0	1	1	0	1	2D	LOADPC(5,0)
0	1	1	0	1	1	0	1	6D	LOADPC(5,1)
1	0	1	0	1	1	0	1	AD	LOADPC(5,2)
0	0	1	1	0	1	0	1	35	LOADPC(6,0)
0	1	1	1	0	1	0	1	75	LOADPC(6,1)
1	0	1	1	0	1	0	1	B5	LOADPC(6,2)

7.2.5 Software to Read PIT's

The subroutine to **READ PIT** counter **0** is reproduced from Section 6.6.3.

```
SUB READPIT0(LOW,HIGH)
    OUT CP,NORM
    OUT CP,SELPC0
    OUT CP,READPC0
    LOW = INP(DP)
    OUT CP,SELPC0
    OUT CP,NORM
    OUT CP,SELPC0
    OUT CP,READPC0
    HIGH = INP(DP)
    OUT CP,SELPC0
    OUT CP,NORM
END SUB
```

When executed, the sequence of control outputs and data inputs carries out the timing pattern in Figure 6.12. HIGH and LOW are returned and the value read is HIGH*256 + LOW. The constants are SELPC0 (**SEL**ect PIT Counter **0**) and READPC0 (**READ PIT** Counter **0**). SELPC0 is the same as in the load sequence. Separate constants and routines read CTR_1 and CTR_2.

The issue here is how to read 16-bit values from any of three counters in any of seven PIT's. Again, declare the constant READPC(I,J) a two-dimensional array.

```
DIM READPC(7,3)
```

SELPC(I,J) is the same as in the load operation. Once values are assigned, the following subroutine reads and returns the HIGH and LOW bytes from any counter of any PIT. Again, I designates the PIT and J the counter.

```
SUB READPIT(I,J,LOW,HIGH)
   OUT CP,NORM
   OUT CP,SELPC(I,J)
   OUT CP,READPC(I,J)
   LOW = INP(DP)
   OUT CP,SELPC(I,J)
   OUT CP,NORM
   OUT CP,SELPC(I,J)
   OUT CP,READPC(I,J)
   HIGH = INP(DP)
   OUT CP,SELPC(I,J)
   OUT CP,NORM
END SUB
```

I and J are specified when the routine is called. For example,

```
CALL READPIT(0,3,LOW,HIGH)
```

causes execution with SELPC(0,3) and READPC(0,3). HIGH*256 + LOW is the 16-bit value from CTR_3 of PIT_0.

READPC(I,J) is the same as SELPC(I,J) except RD* is active. The values are:

| CO_7 | CO_6 | CO_5 | CO_4 | CO_3 | CO_2 | CO_1 | CO_0 | hex | constant |
A_1	A_0	C	B	A	RD*	WR*	1	value	name
0	0	0	0	0	0	1	1	03	READPC(0,0)
0	1	0	0	0	0	1	1	43	READPC(0,1)
1	0	0	0	0	0	1	1	83	READPC(0,2)
0	0	0	0	1	0	1	1	0B	READPC(1,0)
0	1	0	0	1	0	1	1	4B	READPC(1,1)
1	0	0	0	1	0	1	1	8B	READPC(1,2)
0	0	0	1	0	0	1	1	13	READPC(2,0)
0	1	0	1	0	0	1	1	53	READPC(2,1)
1	0	0	1	0	0	1	1	93	READPC(2,2)
0	0	0	1	1	0	1	1	1B	READPC(3,0)
0	1	0	1	1	0	1	1	5B	READPC(3,1)
1	0	0	1	1	0	1	1	9B	READPC(3,2)
0	0	1	0	0	0	1	1	23	READPC(4,0)
0	1	1	0	0	0	1	1	63	READPC(4,1)
1	0	1	0	0	0	1	1	A3	READPC(4,2)
0	0	1	0	1	0	1	1	2B	READPC(5,0)
0	1	1	0	1	0	1	1	6B	READPC(5,1)
1	0	1	0	1	0	1	1	AB	READPC(5,2)
0	0	1	1	0	0	1	1	33	READPC(6,0)
0	1	1	1	0	0	1	1	73	READPC(6,1)
1	0	1	1	0	0	1	1	B3	READPC(6,2)

7.2.6 Summary

The CDI produces 11 control signals for 8253 PIT's. Up to seven chips can be programmed, loaded and read. Three subroutines are needed: PROG-PIT(I,CW), LOADPIT(I,J,LOW,HIGH) and READPIT(I,J,LOW,HIGH). Seventy-seven constants are used: SELPP(I), PROGP(I), SELPC(I,J), LOAD-PIT(I,J) and READPIT(I,J) where I goes from 0 to 6 and J from 0 to 2. Once the subroutines and constants are entered, corrected and saved, they can be used in many applications (and never have to reentered or changed).

The software approach in this book is to have main programs: 1) declare constants and variables integer; 2) dimension arrays; 3) make constants COMMON SHARED; and 4) assign values. Of course, if only one or two PIT's are used, the unneeded constants can be omitted. A complete CDI program shell is given in Section 7.7.3.

7.3 Memory Controls

Section 5.3 describes hardware and software which writes values to and afterward reads values from a succession of addresses in a 6264 memory IC. The purpose of this section is to show how the CDI supports the same operations.

The memory circuit in Figure 5.6 uses five control signals. LD* (LoaD) puts 1's into all bits of four address generating counters. A low-to-high transition of CU (Count Up to the low counter) increments the address by one. Activation of WE* (Write Enable) puts the current value at the data-out port on the bus $d_7...d_0$, and deactivation causes the memory chip to store the value at the current address. A low-to-high of EW (End memory Write) takes data-out off $d_7...d_0$. Activation of OE* (Output Enable) causes the memory to put the value at the current address on $d_7...d_0$ (which software reads through the data-in port).

7.3.1 Memory Hardware

The CDI generates no independent memory controls. Rather, one or several 6264's share CO_7, CO_6, CO_2 and CO_1 with up to seven 8253 PIT's. The specific control-out bits adopted for memory use are given in Figure 7.6 and the following table.

CO_7	CO_6	CO_5	CO_4	CO_3	CO_2	CO_1	CO_0	
A_1	A_0	C	B	A	RD*	WR*	1	PIT connections
CU	LD*	1	1	1	OE*	WE*	1	memory connections

CBA activates a PIT CS* except when equal to 111. So if all control-out bytes in memory operations have CBA = 111, no PIT responds to changes in CO_7, CO_6, CO_2 and CO_1.

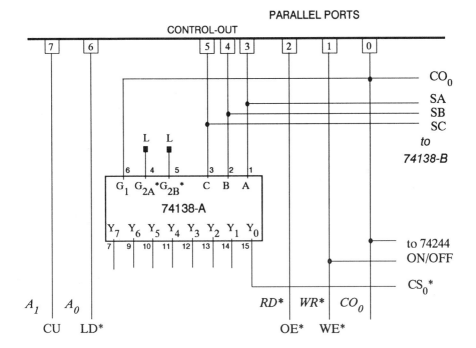

Figure 7.6. Organization of memory controls.

The final hardware issue is management of the data bus $d_7...d_0$ with both 8253's and 6264's connected. Recall from Figure 7.5 that activation of CO_1 (when CO_0 is high) puts the value at the data-out port on $d_7...d_0$. And any CS* going inactive takes the value off. Since the memory's WE* is CO_1, its activation puts the value on. As discussed in Section 5.3.3, data-out values may stay on $d_7...d_0$ until an entire write to memory sequence is finished. Then, the last value is taken off. This can be accomplished by briefly activating CS_0^* while CO_1 (WR*) and CO_2 (RD*) are inactive, thus not interfering with any PIT.

The hardware plan is to wire the 6264 memory and address generator in Figure 5.6 to the CDI's CO_7, CO_6, CO_2, CO_1 and $d_7...d_0$. Also, up to seven 8253 PIT's might be wired to the same lines.

Recall from Section 2.5.1 that a TTL output can be connected to only six or so inputs. The fanout issue is addressed in later chapters when multiple devices are actually wired to the CDI.

7.3.2 Memory Software

In Section 5.4, five subroutines were developed to carry out memory operations. They are reproduced here for the CDI. Six constants are required: NORM (**NORM**al, inactive state), MALOAD (**M**emory **A**ddress **LOAD**), MAADV (**M**emory **A**ddress **ADV**ance), MWRITE (**M**emory **WRITE**), MENDW (**M**emory **END W**rite) and MREAD (**M**emory **READ**). They are reformulated in the table below and, except for MENDW, have the same values as before.

| CO_7 | CO_6 | CO_5 | CO_4 | CO_3 | CO_2 | CO_1 | CO_0 | hex | constant |
CU	LD*	C	B	A	OE*	WE*	1	value	name
1	1	1	1	1	1	1	1	FF	NORM
1	0	1	1	1	1	1	1	BF	MALOAD
0	1	1	1	1	1	1	1	7F	MAADV
1	1	1	1	1	1	0	1	FD	MWRITE
1	1	0	0	0	1	1	1	C7	MENDW
1	1	1	1	1	0	1	1	FB	MREAD

Output of MENDW to the control port activates CS_0*. Output of NORM immediately afterward deactivates CS_0* and causes the data-out port to be taken off $d_7...d_0$.

The five subroutines are identical to those in Section 5.4.1.

```
SUB ZEROADR
    OUT CP,NORM
    OUT CP,MALOAD
    OUT CP,NORM
    OUT CP,MAADV
    OUT CP,NORM
END SUB
```

MALOAD puts 1 in all bits of the address counters. Then MAADV advances the address to zero.

```
SUB ADVADR
    OUT CP,NORM
    OUT CP,MAADV
    OUT CP,NORM
END SUB

SUB READMEM(DVALUE)
    OUT CP,NORM
    OUT CP,MREAD
    DVALUE = INP(DP)
    OUT CP,NORM
END SUB

SUB WRITEMEM(DVALUE)
    OUT CP,NORM
    OUT DP,DVALUE
```

```
      OUT CP,MWRITE
      OUT CP,NORM
END SUB

SUB ENDMEMWR
      OUT CP,NORM
      OUT CP,MENDW
      OUT CP,NORM
END SUB
```

In Chapter 10, four 6264 memory chips, the address generator and a controlling PIT are all wired to the CDI as part of a programmable "fast" data recorder.

7.4 Read ADC Controls

74138-B generates five independent **R**ead **E**nables suitable for obtaining data from ADC's and other devices. Circuitry shown in Figure 7.7 produces RE_2^*, RE_3^*, RE_4^*, RE_5^* and RE_6^*. CO_0 is connected to 74138-B's active low gates G_{2A}^* and G_{2B}^*. G_1 is wired active. CO_5, CO_4 and CO_3 go to the select inputs C, B and A, respectively. To activate RE_i^*, CO_0 must be low and $CO_5CO_4CO_3$ = i.

The absence of RE_0^* and RE_1^* is explained in the next section. Just as in the case of the CS*'s, there is no RE_7^*. Therefore, a control output with CO_0 = 0 (or 1) and $CO_5CO_4CO_3$ = 1 1 1 has no effect on RE*'s (or CS*'s).

The hardware plan is that up to five RE*'s are wired to the active-low read controls of one or several devices. For example, four RE*'s are connected to the HE*'s and LE*'s of two AD573's. Also, $d_7...d_0$ is connected to the devices (as well as to possible PIT and memory IC's).

The question is what sequence of outputs to the control port briefly activates the RE*'s? Circuit connections in Figure 7.7 are summarized in the following table.

CO_7	CO_6	CO_5	CO_4	CO_3	CO_2	CO_1	CO_0	
X	X	C	B	A	X	X	0	RE* connections

As always, the first task is to determine a value for the **NORM**al, inactive state. Since CBA equals 111 activates an unused 74138-B output, NORM could be (using 1 for bits which don't matter):

CO_7	CO_6	CO_5	CO_4	CO_3	CO_2	CO_1	CO_0	hex	constant
X	X	C	B	A	X	X	0	value	name
1	1	1	1	1	1	1	0	FE	NORM

However, it's desirable for NORM to be the same for all software which uses the CDI. What if CO_0 were 1 so NORM equals FF as in previous sections? The consequence is that 74138-A responds to CBA. But CBA = 111 pro-

PARALLEL PORTS

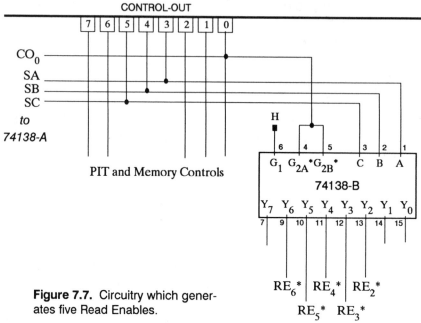

Figure 7.7. Circuitry which generates five Read Enables.

duces no action since Y_7 is not used. Also, $d_7...d_0$ is not affected since CO_1 is high. So the value of NORM can be FF hex.

One of the frustrations in designing hardware/software systems such as the CDI arises from trade-offs. The most reasonable value for NORM is FE hex. Yet, the advantage of NORM having the same value in all CDI operations outweighs this reasonableness.

In addition to NORM, five constants are needed, one for each RE*. Suppose the array **REaD** **D**evice is declared.

        DIM REDD(7)

What value of REDD(I) sent to control-out activates RE_1^*?

CO_7 X	CO_6 X	CO_5 C	CO_4 B	CO_3 A	CO_2 X	CO_1 X	CO_0 0	hex value	constant name
1	1	1	1	1	1	1	1	FF	NORM
1	1	0	1	0	1	1	0	D6	REDD(2)
1	1	0	1	1	1	1	0	DE	REDD(3)
1	1	1	0	0	1	1	0	E6	REDD(4)
1	1	1	0	1	1	1	0	EE	REDD(5)
1	1	1	1	0	1	1	0	F6	REDD(6)

The subroutine **READ D**evice returns an 8-bit value from whatever RE_i^* is connected to.

```
SUB READD(I,DVALUE)
   OUT CP,NORM
   OUT CP,REDD(I)
   DVALUE = INP(DP)
   OUT CP,NORM
END SUB
```

After OUT CP,REDD(I) is executed, the device connected to RE_I^* must put a value on $d_7...d_0$. Then the next statement inputs that value from the data port and assigns it to DVALUE which is returned. For example, after

```
CALL READD(3,Y)
```

Y equals the 8-bit value supplied by the device connected to RE_3^*.

As discussed in earlier chapters, there is a delay between activation of a device's read enable and a value appearing on its data connections. The delay is specified by manufacturers and for an AD573 is 250 ns. If the computer and software are "fast," it's possible for data input to occur too soon. The problem is resolved by outputting a second or third OUT CP,REDD(I).

If more than six RE*s are needed for an application, the four possible values of A_1A_0 could, with additional logic, establish up to twenty-four RE*'s.

7.5 START/STOP*

The final CDI control is Start/Stop* (S/S*). The objective is a signal which can be made high (start) by a unique sequence of outputs to the control port and made low (stop) by a different sequence. The operation must, of course, not interfere with PIT, memory and read device controls. Circuitry is in Figure 7.8.

74138-B's Y_0 (STA*) is connected to the flip-flop's PRESET* input and Y_1 (STO*) is connected to CLEAR*. A brief activation of STA* sets the flip-flop and makes S/S* high (indicating start). A brief activation of STO* clears the flip-flop and makes S/S* low (indicating stop). CO_0 must be low for 74138-B to work. Connections are summarized in the following table.

CO_7	CO_6	CO_5	CO_4	CO_3	CO_2	CO_1	CO_0	
X	X	C	B	A	X	X	0	S/S* control

CBA = 000 activates STA* and CBA = 001 activates STO*.

Start/stop software requires three constants: NORM, STRT (Software sTaRT) and SSTP (Software SToP). NORM is still FF hex (although FE hex is more reasonable as discussed earlier). Values are chosen so that output of STRT activates 74138-B's Y_0 and output of SSTP activates Y_1.

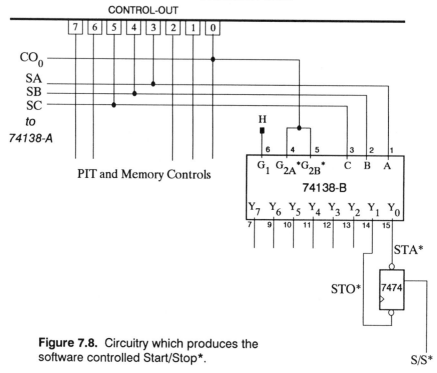

Figure 7.8. Circuitry which produces the
software controlled Start/Stop*.

CO_7	CO_6	CO_5	CO_4	CO_3	CO_2	CO_1	CO_0	hex	constant
X	X	C	B	A	X	X	0	value	name
1	1	1	1	1	1	1	1	FF	NORM
1	1	0	0	0	1	1	0	C6	STRT
1	1	0	0	1	1	1	0	CE	SSTP

CALL START makes S/S* high and CALL SSTOP makes S/S* low.

```
SUB START
    OUT CP,NORM
    OUT CP,STRT
    OUT CP,NORM
END SUB

SUB SSTOP
    OUT CP,NORM
    OUT CP,SSTP
    OUT CP,NORM
END SUB
```

A practice followed so far in this book is never to use the same name for a signal, constant and/or subroutine. This combined with QuickBASIC key words leads to less than ideal names for start/stop hardware and software.

7.6 Monitor Inputs

The CDI sends 10 Monitor lines (M_0...M_9) to the control-in port. They allow software to respond to hardware operations such as data ready. Figure 7.9 shows the relationship between the monitors and control-in bits. CI_0 to CI_6 equal M_0 to M_6, respectively. CI_7 equals M_7 OR M_8 OR M_9. The reason for the OR is to show how to add capacity.

The software operation

```
HSTATUS = INP(CP)
```

gives HSTATUS's 16 bits the following values:

$b_{15}...b_8$	b_7	b_6	b_5	b_4	b_3	b_2	b_1	b_0
all 0's	M_9 OR M_8 OR M_7	M_6	M_5	M_4	M_3	M_2	M_1	M_0

Software polling techniques were presented in Section 3.5.4 and in the example circuits in Chapters 4-6. Additional techniques are illustrated in later chapters.

7.7 CDI Circuit and Software

7.7.1 74244 Buffers and the Complete Circuit

Except for additional buffering, the CDI circuit is given piecemeal in Figures 7.4-7.9. Figure 7.10 is the complete circuit.

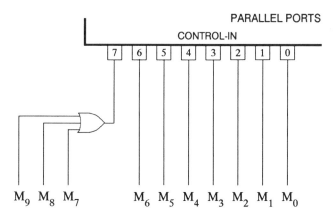

Figure 7.9. Circuitry for Monitor status lines.

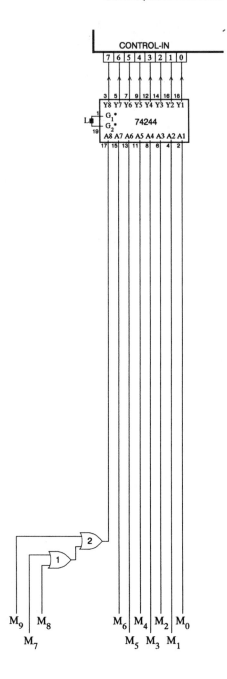

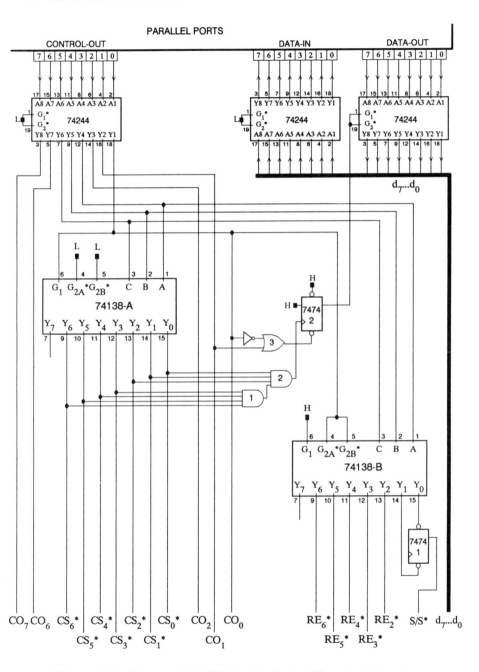

Figure 7.10. The complete CDI circuit with the I/O ports buffered.

Control-in, control-out and data-in are buffered through 74244 IC's. The gates are wired permanently active so the outputs always equal the inputs. Data-out's buffer is controlled as shown in Figure 7.5.

One reason for the buffers is to isolate the parallel ports circuit from the CDI and applications. This keeps any consequence of wiring errors or defective components out of the computer.

Buffering could be accomplished with 74373 IC's. 74244's are used because of electrical characteristics which might be important if "long" cables are needed to connect the CDI to the parallel ports. Also, 74244's have greater fanout and therefore can be connected to more devices (as discussed later when circuits are wired to the CDI).

In addition to the buffers, the OR gate for the high monitor lines is implemented with two 2-input OR's (in a 7432 IC). The 7-input AND in Figure 7.5 is implemented with two 4-input AND's (in a 7421 chip).

Construction details, given in the next section, include specific connectors for the parallel ports and for making the CDI accessible to application circuits.

7.7.2 Hardware Summary

The CDI receives 32 lines from the parallel ports and produces an 8-bit data bus, 18 control signals and 10 monitor inputs. Several control signals serve multiple purposes. The parallel ports lines are:

$CO_7...CO_0$ $CI_7...CI_0$ $DO_7...DO_0$ $DI_7...DI_0$

The CDI makes available to application modules:

$d_7...d_0$

CO_0 CO_1 CO_2 CO_6 CO_7

$CS_0{}^*$ $CS_1{}^*$ $CS_2{}^*$ $CS_3{}^*$ $CS_4{}^*$ $CS_5{}^*$ $CS_6{}^*$

$RE_2{}^*$ $RE_3{}^*$ $RE_4{}^*$ $RE_5{}^*$ $RE_6{}^*$

S/S^*

M_0 M_1 M_2 M_3 M_4 M_5 M_6 M_7 M_8 M_9

Control-out bits have the multiple uses summarized in the following table.

CO_7	CO_6	CO_5	CO_4	CO_3	CO_2	CO_1	CO_0	
A_1	A_0	C	B	A	RD*	WR*	1	PIT controls
CU	LD*	1	1	1	OE*	WE*	1	memory controls (except ENDMEMWR)
X	X	C	B	A	X	X	0	read enables
X	X	C	B	A	X	X	0	start/stop

7.7.3 Software Summary

Throughout this chapter, constants and subroutines were developed mostly for the 8253 PIT, 6264 memory and AD573 ADC. The following program shell contains the BASIC statements needed to call all routines for all cases. Once entered and corrected, the shell can be used over and over.

As discussed in earlier chapters, control signals might not be active long enough if fast computers and compiled languages are used. In the subroutines below, critical signals are lengthened by double and triple outputs.

In the constants below, CP and DP are assigned numbers for the IBM PC/XT/AT ports described in Section 3.2. If different ports or computers are involved, use appropriate values. Also, commercial ports usually require programming. The initialization routines supplied by manufacturers should be added to the shell.

```
'        THE SOFTWARE SHELL
'        Arrays and Constants Defined and Assigned Values
         DEFINT A-Z
         DIM SELPP(7),PROGP(7)
         DIM SELPC(7,3),LOADPC(7,3),READPC(7,3)
         DIM REDD(7)
         COMMON SHARED CP,DP,NORM,STRT,SSTP
         COMMON SHARED SELPP(),PROGP(),SELPC()
         COMMON SHARED LOADPC(),READPC()
         COMMON SHARED MALOAD,MAADV,MREAD,MWRITE,MENDW
         COMMON SHARED REDD()
         CP = &H300 : DP = &H301
         NORM = &HFF
         STRT = &HC6 : SSTP = &HCE
         SELPP(0)  = &HC7
         PROGP(0)  = &HC5
         SELPC(0,0)  = &H07 : SELPC(0,1)  = &H47
         SELPC(0,2)  = &H87
         LOADPC(0,0) = &H05 : LOADPC(0,1) = &H45
         LOADPC(0,2) = &H85
         READPC(0,0) = &H03 : READPC(0,1) = &H43
         READPC(0,2) = &H83
         SELPP(1)  = &HCF
         PROGP(1)  = &HCD
         SELPC(1,0)  = &H0F : SELPC(1,1)  = &H4F
         SELPC(1,2)  = &H8F
         LOADPC(1,0) = &H0D : LOADPC(1,1) = &H4D
         LOADPC(1,2) = &H8D
         READPC(1,0) = &H0B : READPC(1,1) = &H4B
         READPC(1,2) = &H8B
         SELPP(2)  = &HD7
         PROGP(2)  = &HD5
         SELPC(2,0)  = &H17 : SELPC(2,1)  = &H57
```

```
SELPC(2,2) = &H97
LOADPC(2,0) = &H15 : LOADPC(2,1) = &H55
LOADPC(2,2) = &H95
READPC(2,0) = &H13 : READPC(2,1) = &H53
READPC(2,2) = &H93

SELPP(3) = &HDF
PROGP(3) = &HDD
SELPC(3,0) = &H1F : SELPC(3,1) = &H5F
SELPC(3,2) = &H9F
LOADPC(3,0) = &H1D : LOADPC(3,1) = &H5D
LOADPC(3,2) = &H9D
READPC(3,0) = &H1B : READPC(3,1) = &H5B
READPC(3,2) = &H9B

SELPP(4) = &HE7
PROGP(4) = &HE5
SELPC(4,0) = &H27 : SELPC(4,1) = &H67
SELPC(4,2) = &HA7
LOADPC(4,0) = &H25 : LOADPC(4,1) = &H65
LOADPC(4,2) = &HA5
READPC(4,0) = &H23 : READPC(4,1) = &H63
READPC(4,2) = &HA3

SELPP(5) = &HEF
PROGP(5) = &HED
SELPC(5,0) = &H2F : SELPC(5,1) = &H6F
SELPC(5,2) = &HAF
LOADPC(5,0) = &H2D : LOADPC(5,1) = &H6D
LOADPC(5,2) = &HAD
READPC(5,0) = &H2B : READPC(5,1) = &H6B
READPC(5,2) = &HAB

SELPP(6) = &HF7
PROGP(6) = &HF5
SELPC(6,0) = &H37 : SELPC(6,1) = &H77
SELPC(6,2) = &HB7
LOADPC(6,0) = &H35 : LOADPC(6,1) = &H75
LOADPC(6,2) = &HB5
READPC(6,0) = &H33 : READPC(6,1) = &H73
READPC(6,2) = &HB3

MAADV = &H7F : MALOAD = &HBF
MREAD = &HFB : MWRITE = &HFD : MENDW = &HC7
REDD(2) = &HD6 : REDD(3) = &HDE
REDD(4) = &HE6 : REDD(5) = &HEE : REDD(6) = &Hf6
'       APPLICATION PROGRAMS GO HERE

        END


'       The Subroutines

'       Software STOP
        SUB SSTOP
            OUT CP,NORM
```

```
        OUT CP,SSTP
        OUT CP,SSTP
        OUT CP,SSTP
        OUT CP,NORM
    END SUB
'   Software START
    SUB START
        OUT CP,NORM
        OUT CP,STRT
        OUT CP,STRT
        OUT CP,STRT
        OUT CP,NORM
    END SUB
'   PROGram PIT I with CW
    SUB PROGPIT(I,CW)
        OUT CP,NORM
        OUT CP,SELPP(I)
        OUT DP,CW
        OUT CP,PROGP(I)
        OUT CP PROGP(I)
        OUT CP,SELPP(I)
        OUT CP NORM
    END SUB
'   LOAD PIT I, Counter J with LOW, HIGH
    SUB LOADPIT(I,J,LOW,HIGH)
        OUT CP,NORM
        OUT CP,SELPC(I,J)
        OUT DP,LOW
        OUT CP,LOADPC(I,J)
        OUT CP,LOADPC(I,J)
        OUT CP,SELPC(I,J)
        OUT CP,NORM
        OUT CP,SELPC(I,J)
        OUT DP,HIGH
        OUT CP,LOADPC(I,J)
        OUT CP,LOADPC(I,J)
        OUT CP,SELPC(I,J)
        OUT CP,NORM
    END SUB
'   READ PIT I, Counter J, Return LOW, HIGH
    SUB READPIT(I,J,LOW,HIGH)
        OUT CP,NORM
        OUT CP,SELPC(I,J)
        OUT CP,READPC(I,J)
        OUT CP,READPC(I,J)
        LOW = INP(DP)
        OUT CP,SELPC(I,J)
        OUT CP,NORM
        OUT CP,SELPC(I,J)
```

```
            OUT CP,READPC(I,J)
            OUT CP,READPC(I,J)
            HIGH = INP(DP)
            OUT CP,SELPC(I,J)
            OUT CP,NORM
      END SUB
```

' **READ ADC** Type Device I, Return DVALUE
```
      SUB READDD(I,DVALUE)
            OUT CP,NORM
            OUT CP,REDD(I)
            OUT CP,REDD(I)
            DVALUE=INP(DP)
            OUT CP,NORM
      END SUB
```

' **ZERO** Memory **AD**d**R**ess
```
      SUB ZEROADR
            OUT CP,NORM
            OUT CP,MALOAD
            OUT CP,NORM
            OUT CP,MAADV
            OUT CP,NORM
      END SUB
```

' **AD**Vance Memory **AD**d**R**ess
```
      SUB ADVADR
            OUT CP,NORM
            OUT CP,MAADV
            OUT CP,NORM
      END SUB
```

' **READ** Current **MEM**ory Address, Return DVALUE
```
      SUB READMEM(DVALUE)
            OUT CP,NORM
            OUT CP,MREAD
            OUT CP,MREAD
            VALUE = INP(DP)
            OUT CP,NORM
      END SUB
```

' **WRITE** DVALUE to Current **MEM**ory Address
```
      SUB WRITEMEM(DVALUE)
            OUT CP,NORM
            OUT DP,VALUE
            OUT CP,MWRITE
            OUT CP,MWRITE
            OUT CP,NORM
      END SUB
```

' **END** a **MEM**ory **WR**ite Sequence
```
      SUB ENDMEMWR
            OUT CP,NORM
            OUT CP,MENDW
```

```
      OUT CP,MENDW
      OUT CP,NORM
   END SUB
```

7.8 Construction

Except for the IBM and Apple parallel ports described in Appendices D and E, all circuits so far in this book have been constructed on breadboards. However, a more organized and permanent method is needed for the CDI module. This section describes building the circuit on a wire wrap card. Specific recommendations are made for every detail.

The description is based on experience from constructing the CDI several times. Other sets of details are certainly possible.

7.8.1 Components

The CDI circuit in Figure 7.10 requires 10 integrated circuits.
(1)   Four 74LS244's to buffer the parallel ports.
(2)   Two 74LS138's to set up the CS*'s and RE*'s.
(3)   A 74LS74 with two flip-flops, one to control the data bus and one to set up S/S*.
(4)   A 74LS32 OR with one gate for bus control and two for combining M_7, M_8 and M_9.
(5)   A 74LS21 4-input AND so any CS* can take data-out off $d_7...d_0$.
(6)   A 74LS04 INVERTER with one gate involved in bus control.

In addition to the IC's, two connectors are required.
(1)   One receives the parallel ports cable. For the IBM and Apple ports, a 34-pin protected male wire wrap connector (Amphenol type 842-816-3427-035) takes the ribbon cable described in Appendices D and E. Pin assignments are reproduced in Figure 7.11. If commercial ports are used, then an appropriate wire wrap connector must be substituted and different pin assignments followed.

34	33	32	31	30	29	28	27	26	25	24	23	22	21	20	19	18

CI_7 CI_5 CI_3 CI_1 CO_7 CO_5 CO_3 CO_1 G DI_7 DI_5 DI_3 DI_1 DO_7 DO_5 DO_3 DO_1

PARALLEL PORTS CONNECTOR

CI_6 CI_4 CI_2 CI_0 CO_6 CO_4 CO_2 CO_0 G DI_6 DI_4 DI_2 DI_0 DO_6 DO_4 DO_2 DO_0

1	2	3	4	5	6	7	8	9	10	11	12	13	14	15	16	17

Figure 7.11. Pin assignments for the IBM PC and Apple parallel ports connector described in Chapter 3 and Appendices D and E.

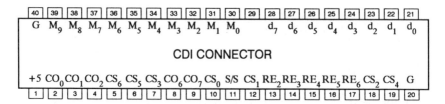

Figure 7.12. Pin assignments for the connector between the CDI and applications modules. (The "*" on active low signals is not included in the labels.)

(2) The CDI is connected to applications modules by a 40-conductor ribbon cable with dip plugs on both ends. Zero insertion pressure sockets and receptacles are recommended for the plugs (3M/Textool type 240-3346 and 240-3599). The pin assignments, shown in Figure 7.12, are:

pins	signals (respectively)
2, 3, 4, 8, 9	CO_0, CO_1, CO_2, CO_6, CO_7
10, 12, 18, 7, 19, 6, 5	CS_0*, CS_1*, CS_2*, CS_3*, CS_4*, CS_5*, CS_6*
11	S/S*
13, 14, 15, 16, 17	RE_2*, RE_3*, RE_4* RE_5*, RE_6*
28...21	$d_7...d_0$
39...30	$M_9...M_0$
1	+5
20, 40	ground

In addition to IC's and connectors, these components are needed.
(1) A wire wrap card. The 6 1/2 by 4 1/2 inch Vector Model 8007 or equivalent is recommended.
(2) Four 20-pin wire wrap sockets.
(3) Four 14-pin sockets.
(4) Two 16-pin sockets.
(5) A wire wrap tool and several spools of different color wire.
(6) A packet of individual wire wrap terminals.
(7) Eight to ten .01 μf disc capacitors.
    Vendors for all components are listed in Appendix B.

7.8.2 Layout

Figure 7.13 gives the circuit layout. Solder and/or epoxy the 10 IC and 2 connector sockets to the card approximately as shown.

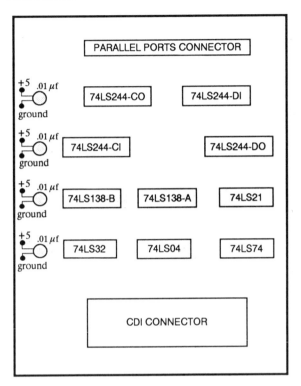

Figure 7.13. CDI circuit layout on a wire wrap card.

7.8.3 +5 and Ground Connections

As shown in Figures 7.11 and 7.12, ground pins on the parallel ports connector are 9 and 26 and on the CDI 20 and 40. +5 volt power for the circuit comes from the CDI's pin 1.

Solder pairs of individual wire wrap terminals (one for +5 and the other for ground) adjacent to each row of IC's as shown in Figure 7.13. Also, solder .01 µf noise capacitors between the terminals. With red wire, attach pin 1 of the CDI connector to the bottom +5 terminal and then upward to the top +5 terminal. With black wire, connect CDI pin 20 to pin 40, then pin 40 to the bottom ground terminal, upward to the top terminal, and finally to ports connector pins 9 and 26.

Wire wrapping itself is straightforward. Just practice and observe.

The next step is to wire the IC sockets. A series of "wire wrap" drawings, the first in Figure 7.14, facilitate the work. They show the 2 connectors and 10 IC's from above so all pin 1's are on the bottom left. Designated bold

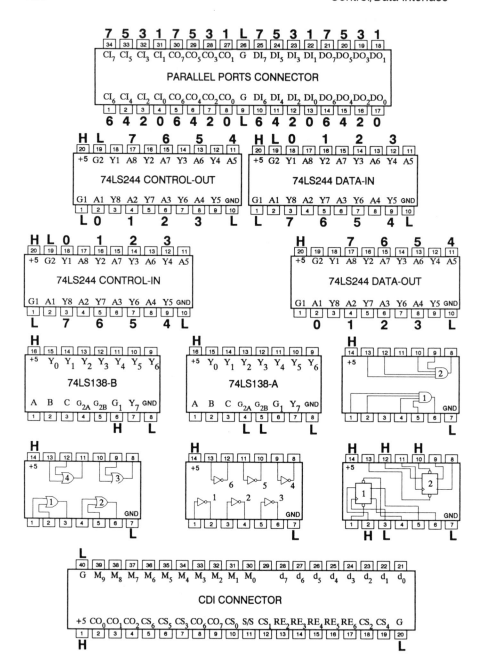

Figure 7.14. Wire wrap drawing showing connections for +5 and ground and for the parallel ports connector.

characters are placed above or below pins to be connected. Use Figure 7.14 to wire +5 to each **H** labeled pin and ground to each **L** pin. Start on the left with the separate +5 and ground terminals and wire across a row of IC's stopping on the right. It's important not to make ground "loops." (Ignore for now the digit characters in the figure.)

Another approach is to use drawings similar to Figure 7.14 but "upside-down-and-backward." This is because the wire wrap card has to be turned over to actually make connections. The "right-side-up" method here encourages thinking about each step in terms of what's being wired. And it facilitates testing and troubleshooting described in Section 7.9.

7.8.4 Wiring the 74244's

In Figure 7.14, each data-out pin on both the ports connector and data-out 74244 is labeled with a digit corresponding to the bit number. Wire **0** to **0**, **1** to **1**, etc. Do the same for data-in, control-out and control-in. Since the same digits appear four times on the ports connector, be sure the DO's, DI's, CO's and CI's go to the right 74244. Note that 74244 inputs are A's and outputs Y's. Also note that 74244 bit numbers go from 1 to 8 and the ports connector from 0 to 7.

Outputs from the data-out 74244 go to inputs of the data-in 74244 and to the CDI connector. Use Figure 7.15 to wire **0** to **0** to **0**, **1** to **1** to **1**, etc.

Inputs from the control-in 74244 are wired to the CDI connector and OR-2's output. Use Figure 7.15 to wire **A** to **A**, **B** to **B**, ..., and **H** to **H**.

The control-out 74244 is more complex. For instance, the circuit in Figure 7.10 shows that Y1 (CO_0) goes to both 74138's, INVERTER-4 and the CDI connector. These pins are all labeled **S** in Figure 7.15. Everything Y2 is connected to is labeled **T**. Similarly, the Y3 to Y8 connections are labeled **U**, **V**, **W**, **X**, **Y** and **Z**, respectively. Wire all **S**'s, all **T**'s, etc.

It's important not to work blindly. Always refer to the actual circuit diagram as each step is made.

7.8.5 Wiring the Rest of the Circuit

Figure 7.16 shows the remaining connections.

Wire 74138-A's outputs, labeled **0** to **6**, to the AND gates and CDI connector.

Wire 74138-B's outputs, **A** to **G**, to flip-flop-1 and the CDI connector. Wire flip-flop-1's output **M** to the CDI connector.

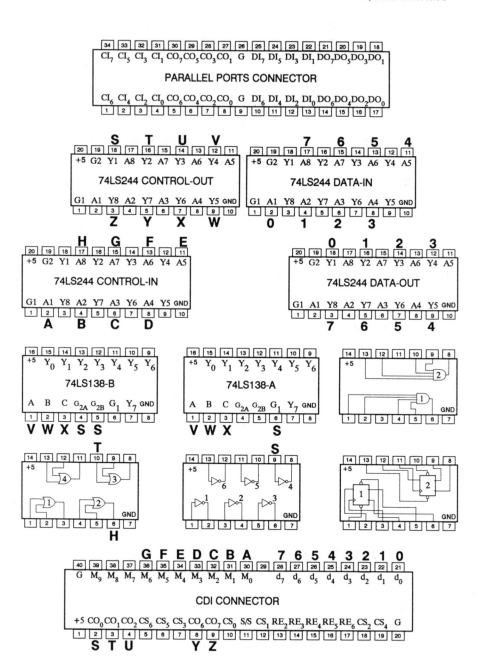

Figure 7.15. Wire wrap drawing showing connections to the two 74138's.

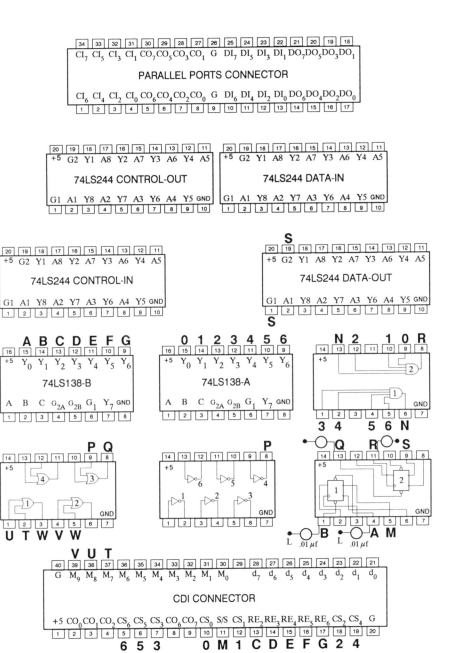

Figure 7.16. Wire wrap drawing showing the remaining CDI circuit connections.

Wire INVERTER-4's output **P** to OR-3's input. Wire OR-3 and AND 2 outputs, **Q** and **R**, to flip-flop-2. Wire flip-flop-2's output **S** to both gates of the data-out 74244.

Finally, wire the CDI connector's M_7, M_8 and M_9, labeled **T**, **U** and **V**, to OR-1 and OR-2 as shown. Wire OR-1 output **W** to an OR-2 input. (OR-2's output was previously connected to control-in.)

2.8.6 Noise Capacitors

Experience has shown that both flip-flop-1 (S/S*) and flip-flop-2 (data-out control) can be inadvertently set and/or cleared. To protect against the possibility, connect .01 μf noise capacitors between F-F-1's SET* and ground and CLEAR* and ground, and between F-F-2's CLOCK and ground and CLEAR* and ground. If the wire wrap card has a grounded metal "grid," just solder the capacitors between IC socket wire wrap pins and the grid. Otherwise, solder individual wire wrap terminals to the card, wire them to ground, and solder the capacitors between the IC sockets and terminals. In either case, the capacitors should be on the components side of the card.

This completes construction. Visually inspect the work. Install the IC's.

7.9 Testing and Troubleshooting

In order to test the circuit and software, run a 40-conductor ribbon cable from the CDI connector to a breadboard. A good idea is to install a wire wrap socket on the breadboard for the cable's dip plug. Pin assignments are the same as in Figure 7.12. Connect the breadboard system's +5 to pin 1 and ground to pins 20 and 40. The tools required for the tests below are a logic probe, IC test clips, a multimeter (which measures resistance and DC voltage) and an oscilloscope. It's assumed the parallel ports and cables work as specified in Chapter 3.

7.9.1 Continuity Verification

With the power off and parallel ports not connected, test the wiring by confirming that the resistance between pins which should be connected is zero. Put the multimeter probes directly on IC pins, parallel ports connector pins and CDI socket pins (on the breadboard). Follow the order used in construction. Figures 7.14-7.16 are helpful in locating pins. Common errors are mixing up the order of a set of bits (such as $CO_7...CO_0$) and omitting connections.

7.9.2 Test +5 and Ground

Turn the breadboard power supply on. Measure the voltage at all four +5 terminals on the CDI card. If it's not greater than 4, disconnect the source from the CDI and measure directly. If the value is now reasonable, the circuit is loading the supply either through a short or defective IC or because the supply is inadequate. Try a different voltage source. Replace all chips. If the problem remains, repeat the continuity tests.

After the voltage is correct, use Figure 7.14 to check all **H** and **L** connections with the logic probe.

7.9.3 Enter and Proof the Software Shell

Enter, save and print the software shell in Section 7.7.3. Proof the print out and make corrections. Just as in testing continuity, the more carefully this is done, the better.

7.9.4 Test PIT Controls

The goal is to confirm that each subroutine produces the desired control sequence. The method is to call routines in an infinite loop and observe relevant signals on an oscilloscope.

Attach the parallel ports cable to the CDI. Insert the following at the point in the software shell where application programs go:

```
REPT:  CALL PROGPIT(0,0)
       GOTO REPT
       END
```

Save with an appropriate filename and then run. Observe CS_0^* on oscilloscope channel 1. (A good place to connect the scope is at the end of the CDI cable on the breadboard.) The signal should briefly activate once each loop and produce a stable scope pattern.

Recall that a QuickBASIC infinite loop is aborted by typing CONTROL-BREAK.

If CS_0^* is wrong, immediately check CO_1 (WR*) on channel 2 which during PROGPIT should activate while CS_0^* is active. If WR* is also wrong: 1) test the parallel ports while they are connected to the CDI circuit; 2) repeat the continuity checks on 74138-A; and 3) recheck the PROGPIT subroutine and relevant constants. If the problem is not found, stop and check one of the circuits in Chapters 4-6. At this point, it's important to get something to work. Before going back to the CDI, review the troubleshooting techniques in Sections 4.7, 5.5 and 6.7.

After CS_0^* and WR* check out, with the program running verify that RD*, A_1 and A_0 stay high. Next, test the data bus control circuit by looking at CS_0^* on channel 1 and flip-flip-2's output via an IC test clip on channel 2. The latter should go low after CS_0^* (when WR* goes active) and not return high until CS_0^* does. If the signal is wrong, check the AND, INVERTER, OR and flip-flop wiring. Software cannot cause the problem. Finally, when the data-out 74244 is on, check that each bit of $d_7...d_0$ is zero. A good place for the scope probe is the CDI connector on the breadboard.

A judgment is necessary at this point. Should the data-out bits be checked for wiring reversals. This could be done by calling the subroutine with CW's such as 1, 2, 4, 8, etc. However, in view of the continuity checks, it's probably not necessary.

Finally, during PROGPIT and LOADPIT, WR* should be active at least 400 ns (as discussed in Section 6.4.2). Measure the signal's length and extend as necessary.

Next, change the program to generate CS_1* and repeat the tests above. Do the same for the remaining CS*'s.

If PROGPIT works, the hardware for LOADPIT must also work. Test the software with a procedure similar to the one above.

Finally, check the READPIT hardware and software. As discussed in Sections 6.6.4 and 6.7.3, RD* should be active at least 600 ns. Measure and extend as necessary.

Great patience and discipline are needed for these tests. For instance, suppose one of the chip selects was wired to the wrong AND pin but both its continuity and software checks were skipped. Then, years later that CS is finally used in an application circuit. Naturally, no one would think the CDI is the problem.*

7.9.5 Test Memory Controls

Memory uses the same CDI hardware as the PIT's. Therefore, it's only necessary to test the software. Put each subroutine in an infinite loop and observe the relevant controls on an oscilloscope. Follow suggestions in Sections 5.5.3 and 5.5.4 on cirtical times.

7.9.6 Test Read Enables

Replace the main program loop with:

```
REPT:  CALL READD(2,X)
       GOTO REPT
       END
```

Observe RE_2* at the CDI connector on the breadboard. It should briefly activate each loop. If not, recheck 741348-B's wiring and recheck the routine and constants. Test the remaining RE*'s.

For an AD573, the read enables HE* and LE* must be active at least 250 ns. Measure and extend the RE*'s as necessary. Also, when RE*'s are used with other devices, be sure timing specifications are met.

7.9.7 Test S/S*

Replace the loop with:

```
REPT:   CALL START
        CALL SSTOP
        GOTO REPT
        END
```

Observe S/S* on an oscilloscope. It should repeatedly alternate between high and low. If S/S* is wrong, check the relevant hardware and software. If the computer is especially "fast" and/or the program compiled, the noise capacitors on flip-flop-1's PRESET* and CLEAR* may "over" filter the signals. Look at both on the scope. If they don't reach zero volts, lengthen the activation by adding output statements to the subroutines.

The following procedure tests some of the ways the flip-flops might be inadvertently set and/or cleared. Replace the loop with the following more complex main program.

```
REPT:   CALL START
        CALL PROGPIT(0,0)
        CALL READD(2,X)
        CALL SSTOP
        GOTO REPT
        END
```

Put S/S* on scope channel 1. Then, check the following signals on channel 2:
(1) CS_0* should briefly activate shortly after S/S* goes high. Be sure S/S* is not affected.
(2) RE_2* should activate a bit later as S/S* stays high.
(3)   The data-out 74244's gates should go low after CS_0* and return high with CS_0*. Check that the gates don't activate at any other time.

7.9.8 Test the Monitor Lines

The final test is the monitor lines. Enter this application program in the software shell.

```
REPT:   STATUS = INP(CP)
        PRINT STATUS
        INPUT "Any key to repeat ";A$
        GOTO REPT
        END
```

Wire M_0 (only) low on the breadboard CDI connector and run the program. STATUS should be 254 since the disconnected bits are seen as high. Wire

M_1 low and repeat. See Section 3.5.2 for the expected decimal value. Continue through M_6. Finally, wire M_7, M_8 and M_9 all low and check that the input is the expected 127. If any monitor bit is wrong, recheck the hardware.

7.9.9 Save for Future Use

Save the various CDI test programs in case the circuit has problems later on or another CDI is built.

Exercises

1. Sketch two 8253's and the memory circuit (a 6264 and the address counters) all wired to the CDI connector so subroutines and constants in the software shell control the devices.

2. Sketch one 8253 and two AD573's wired to the CDI connector so PIT and RE* software operate the devices.

3. What would have to be added to the CDI circuit so the RE*'s (if not needed to read devices) might serve as CS*'s so more than seven 8253's can be interfaced? Be careful with CO_0 and be sure the original CS*'s still work.

4. Show how CO_6, CO_7 and another 74138 can expand the number of RE*'s. Give software including constants which activate the RE*'s. What effect does the circuit addition have on PIT's and the memory circuit?

5. Redo the scaler example in Section 6.8 with two 8253's wired to the CDI. Program the extra counter(s) to replace the breadboard system pulser to trigger the 1.0000 second interval. Initially, set up the system so counting occurs for 1.0000 second once every 10.000 seconds. This leaves 9 seconds to read the count value before next interval. Make the total data acquisition time selectable. Consider using start/stop. Be sure the monitor software works. Draw the circuit connected to the CDI and write the applications program. If an additional array is needed, it must be dimensioned along with the existing arrays in the software shell.

6. Devise hardware and software so an AD573 determines voltage 100 times a second a selectable number of times. The circuit should consist of one or two 8253's and an AD573 wired to the CDI. Use a 1.0000 MHz digital clock and PIT's for timing control. Unlike Chapter 4, have

a PIT counter generate the CONVERT pulse. The monitor operation still requires a flip-flop as discussed in Section 4.4. Data values should be saved in an array (dimensioned with the other arrays in the software shell). Sketch the circuit and write the applications program.

7. Use a PIT counter perhaps in another 8253 so hardware controls the total number of readings in Exercise 6. The input parameter should be the data acquisition time (e.g. 10 seconds means 1000 readings). Use another monitor input to determine when overall acquisition is done.

8. Devise a system with one output which is software selectable from among clock signals of 1 MHz, 100 KHz, 10 KHz, 1000 Hz, 100 Hz, 10 Hz and 1 Hz. Use six counters in two 8253's and the 1.0000 MHz digital oscillator to generate the clocks. A major problem is circuitry which allows software through the control port to select the output. Recall the 74151 data selector/multiplexer discussed in Section 2.8. Since no RE*'s and memory are involved, CO_1, CO_2 and CO_6 might serve as 74151 selects. Give both the circuit connected to the CDI and the applications program. Carefully define control signals and output byte constants.

9. The system in Exercise 8 is more interesting if the frequency is selectable for as many values as possible, say, between 1 HZ and 1 MHz. Since a PIT counter can be programmed to divide an input clock by 2, 3, 4, ... 65535, what are the possibilities? Suppose a 2.0000 MHz clock is available (the highest frequency recommended for an 8253). After reviewing PIT Mode 2 from Section 6.2.2, write software for the best case.

10. Interface the switch encoder in Figure 2.50 so that whenever counting freezes the active switch number is input. The difficulty is how software determines when a switch is active and how not to miss a briefly active switch. An approach is to feed CTRL's inverse to a 7474's CLOCK input. Freezing sets or clears the flip-flop which software monitors. After inputting data, software clears or sets the flip-flop and counting resumes. Sketch the hardware wired to the CDI and write the applications program.

11. Revise the data recorder circuit in Figure 5.11 so one or two 8253's allow the rate of recording to be selectable over a range from 10 Hz to 1 MHz and allow the number of recordings to be selectable. Assume a 2.0000 MHz digital clock is available. Sketch the circuit interfaced to the CDI and write the application program.

12. Revise the digital signal generator in Chapter 5's Exercise 8 so two 8253's make selectable: 1) the rate at which stored values are "played" back

(from 10 Hz to 1 MHz); 2) the number of values played (from 2 to 8192); and 3) the number of times the sequence is repeated (from 1 to 64K). Sketch the circuit interfaced with the CDI and write the application program.

13. Build, test and troubleshoot the CDI. Then, do a couple of the exercises above.

8 ADC Module

The Analog Devices AD573 10-bit, 20 μsec analog-to-digital converter was selected for the data acquisition systems in this book. In the PDC plan, two of the IC's comprise the ADC module which interfaces to whatever application circuits need voltage measurements. Reasons for picking this particular device are in Chapter 4.

Obviously, more than two ADC's are necessary for conceivable applications. And cases may arise where faster and/or more accurate measurements are required. On the other hand, slower, less accurate and less expensive converters might be desirable when many voltages are measured. Because the devices are in a separate module, additional AD573's or some number of different ADC's can be substituted with minimal changes in the rest of the system.

> ADC's such as the ADC0808 (Exercise 6 in Chapter 4) measure voltage from up to eight sources. Select controls determine which input is converted. However, the plan here is for two separate IC's to measure two voltages. This ensures simultaneous conversion of both channels, simplifies control requirements, and minimizes the few millivolts of cross-talk common in multiplexed chips. (Cross-talk means the voltage in one channel influences converted values in another.)

In addition to hosting conversion devices, the ADC module receives power supply voltages and sends +5 and ground to the rest of the PDC (as shown in Figure 7.1). The scheme allows the -12 volts required for AD573's to go only to the module where it's needed.

8.1 Circuit

The AD573 is a 20-pin IC. Ten connections are Data outputs (D0...D9). Three are control inputs: CV (ConVert), HE* (High byte Enable) and LE* (Low byte Enable). One is the control out DR* (Data Ready). Four are for power and ground: +5 volts, digital ground, -12 volts and analog ground. Another is for the voltage to be measured. And the last, BO (Bipolar-Offset), selects the input voltage range. Operation of the chip was covered in Chapter 4.

The ADC module contains two converters, labeled A and B. Communication with applications circuits requires the following connections: CV (to both devices for simultaneous conversion); HE-A*, LE-A*, HE-B* and LE-B* (which command the chips to make data available); the CDI data bus $d_7...d_0$ (over which converted values are input by the computer); AD573-A's DR* (DR* from AD573-B is identical); +5 volts (for the application and CDI modules); and ground. The connections are made with 16-conductor ribbon cable with dip plugs on both ends. Sockets for the plugs (henceforth called the ADC connector) are on both the CDI and application modules. Figure 8.1 shows the two ADC's wired to the connector. The power supply and ground setup is the same as in Figure 4.2. D0 and D2 are both wired to

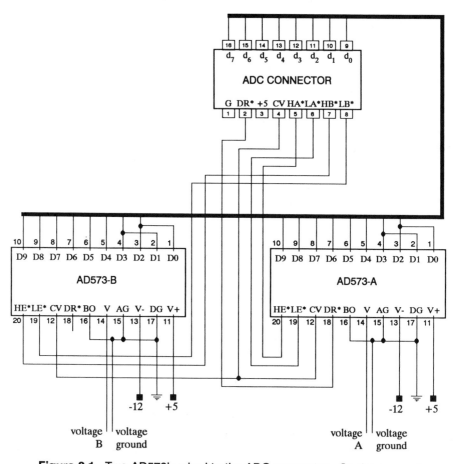

Figure 8.1. Two AD573's wired to the ADC connector. On the connector pin assignments, HA* is HE* for AD573-A, etc.

d_0, and D1 and D3 are wired to d_1 (also the same as in Chapter 4). BO is grounded making the input voltage range 0 to 10.

8.2 Layout and Components

The module is constructed on a wire wrap card with the layout in Figure 8.2. Components are:

(1) A 6 1/2 by 4 1/2 inch wire wrap card with ground plane, such as Vector Model 8007.
(2) Two AD573JN IC's (JN is the sub-model type).
(3) Five 10 μf (or so) tantalum capacitors.
(4) A modular phone jack for power supply connections (preferably panel mount type 623K).
(5) A 16-pin wire wrap socket (for the ribbon cable) such as 3M/Textool zero insertion socket (model 216-3340) and receptacle (216-3593).
(6) Two female BNC connectors.

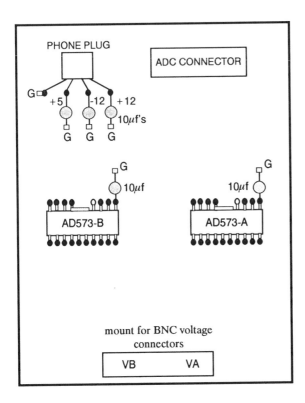

Figure 8.2. Wire wrap card layout for the ADC module (from above the components side).

(7)   Short strips of shielded cable to run from the AD573 voltage inputs to the BNC's.
(8)   A rectangular piece of Plexiglas to support the BNC's.
(9)   A bag of individual wire wrap terminals.
(10) A wire wrap tool and different color wire.
      Vendors for all components are given in Appendix B.

Many other possibilities exist for specific components. For instance, shielded cables and BNC connectors are unnecessary for most situations. The purpose here is to illustrate one complete case.

8.3 Construction

There are two approaches to construction of an ADC module. One is to conservatively follow manufacturer's recommendations and solder the AD573's to the wire wrap card. The other is to use 20-pin wire wrap sockets for the IC's. The former is tedious and makes replacement of a failed chip difficult. The latter results in greater potential for noise. The following details are for the solder approach.

8.3.1 Hardware

Epoxy the phone jack on the ground plane side of the wire wrap card approximately as shown in Figure 8.2. Solder or epoxy the receptacle for the 16-pin zero insertion socket. Screw the socket to the receptacle.

Place the ADC chips on the ground plane (component) side approximately as shown. Then put separate wire wrap terminals in the rows of holes just outside pins 1-10, 11-14 and 17-20 on both chips. Remove the chips and solder the terminals on the wiring (non-ground plane) side of the card. Remount the IC's and carefully solder to the card (on the wiring side). Then, create on the wiring side solder bridges between the corresponding chip and wire wrap terminals. Next, make solder bridges between pins 15, 16 and 17 of both chips. Finally, solder pin 17 (digital ground) to the ground plane on the components side.

For both AD573's, solder $10 \mu f$ (or so) tantalum capacitors between pin 11 (+5 volts) and the ground plane as recommended by the manufacturer.

Solder four wire wrap terminals adjacent to the phone jack as shown. On the component side, solder the black phone cable (ground) to the left terminal, the red cable (+5) to the next terminal, the green (-12) to the next and the yellow (+12) to the right terminal. Solder $10 \mu f$ (or so) tantalum capacitors between +5 and the ground plane, between +12 and the plane and between the ground plane and -12. Finally, make a solder bridge between the ground (left) power supply terminal and the plane.

Drill holes for the BNC connectors in the Plexiglas. Mount the BNC's. Epoxy (or otherwise firmly attach) the Plexiglas to the wire wrap card.

8.3.2 Wiring

Figure 8.3 shows the ADC circuit in the wire wrap drawing format from Chapter 7.

Run a shielded cable between the right BNC and AD573-A's pin 14. Repeat for AD573-B. For each cable, solder the ground shield to the BNC and to the wire wrap terminal adjacent to ADC pin 17.

Wire wrap together all IC socket pins and terminals labeled **L** (ground) and all pins and terminals labeled **H** (+5). Wire the -12 volt phone jack terminal (**Z**) to pin 13 of both ADC's.

The data bus $d_7...d_0$ goes from the ADC connector to both chips. Wire **0**'s, **1**'s. etc.

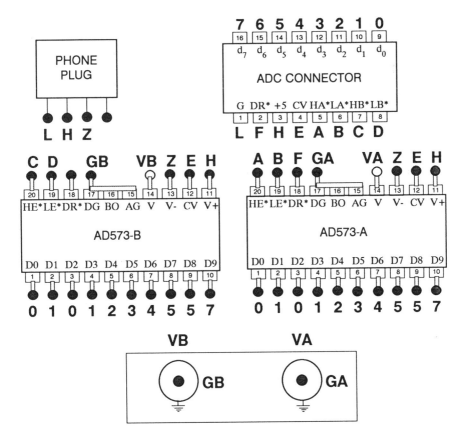

Figure 8.3. Wire wrap drawing for the ADC module (component side).

Wire HE-A*, LE-A*, HE-B* and LE-B* between the respective ADC's and connector. Specifically, connect **A** to **A**, **B** to **B**, **C** to **C** and **D** to **D**. Wire CV (**E**) from the connector to both chips. Finally, wire DR* (**F**) from AD573-A to the connector.

8.3.3 Power Supply and Card Rack

In order to test the ADC module, +5 and -12 volts must be supplied. Also, at some point, the various wire wrap cards must be organized and connected. An approach is to use a card rack for both the modules and power supply. The H. H. Smith Model 3050 universal printed circuit rack is recommended. An inexpensive switching power supply is sufficient. The Jameco Model FCS604A fits in the rack and supplies +5 volts at 5 amperes, and +12, -12 and -5 volts all at 1 amp. Finally, a standard telephone cable runs from the supply to the power jack on the ADC module. Connect the red phone lead to the supply's +5, black to ground, yellow to +12 and green to -12.

Again, many other rack, power supply and cabling setups are possible. The details here illustrate one approach.

8.4 Testing and Troubleshooting

The following procedures test the ADC module apart from an application circuit. The tools required are a breadboard system, a multimeter which measures resistance and voltage, a logic probe and an oscilloscope.

First, visually inspect the card for wiring and soldering problems. Then, with a multimeter set to measure resistance, check all wiring by verifying that zero resistance exists between pins which should be connected. Also, check that no pins are incorrectly shorted to ground.

Connect the power supply and measure +5 and -12 volts at all relevant pins on both ADC's. If a value is wrong, check the voltage with the supply disconnected from the module. Look for shorts and repeat the continuity tests if necessary.

Next, connect the ADC module to a breadboard with a 16-conductor ribbon cable. A wire wrap socket is recommended for the breadboard end of the cable. The following two procedures should reveal incorrect wiring or defective ADC's.

Connect a breadboard system clock set on 10 KHz to CV (ADC connector pin 4). Put the clock on oscilloscope channel 1 and DR* (pin 2) on channel 2. DR* should go high shortly after CV returns low and remain high 20 or so μsec. Also, check DR* from AD573-B. If either is wrong, check CV as it reaches the ADC's. If CV is right and a DR* wrong, there is either a wiring error or defective ADC. Redo the continuity and voltage tests above.

For the second test, connect breadboard system logic switches, set on high, to the read enables (HE-A*, etc.). Wire an active high pulser to CV. Finally, make both voltage inputs zero (by shorting to ground). Activate the pulser. Since conversion takes 20 μsec, it's over essentially as soon as the pulser is released. Activate HE-A* and check d_7...d_0 with the logic probe. All bits should be low. Deactivate HE-A* and activate LE-A* and check d_0 and d_1. They should be low but it's possible noise could make the low bit high. d_2...d_7 should be in a Hi-Z state and produce blanks on the probe. Similarly check AD573-B. Next, connect a known DC voltage, perhaps from a battery, to the ADC's. Repeat the tests except this time several bits should be high. Noise might be investigated by looking repeatedly at d_0 after conversion. If it's always the same, noise is small. Fluctuations in d_1 indicate noise of more than 10 mv. This test does not reveal possible wiring reversals in d_7...d_0. However, the earlier continuity checks should pick those up.

When the ADC module passes all these tests, it should work properly with application circuits.

8.5 Conclusions

The three support components for the PDC are now complete. The parallel ports, CDI (including software) and ADC's are available for a variety of data acquisition systems. The next two chapters illustrate techniques for voltage, time and counting measurements as well as how to use the support modules.

Exercises

1. Suppose an application requires four AD573's but only at 8-bit accuracy. Sketch an ADC module which meets this need with no changes in the 16-conductor ribbon cable.

2. Design an ADC module for the ADC0808 in Chapter 4's Exercise 6 so eight voltages can be converted and the values read. Include the 1.0000 MHz clock in the module. If necessary, go to a 24-conductor cable and specify pin assignments. Do not include the flip-flop for monitoring data ready. Consult manufacturer's specification as necessary.

3. Devise an application circuit which uses the CDI and ADC modules to replicate the voltmeter example in Section 4.6 except voltage from two sources are read and displayed. The software generated CV must now work through the CDI. Sketch the circuit and write the software. Con-

sider using the CDI's read device routine. Assume no PIT's or memories are present.

4.   Devise an application circuit so the instrument in Chapter 7's Exercise 6 works with the ADC and CDI modules. Sketch the circuit and list software differences, if any.

5.   Build, test and troubleshoot the ADC module. Then, do Exercise 3 or 4. Use one or several breadboards connected by ribbon cable to the ADC and CDI modules.

9 Comprehensive Measurement System

9.1 Overview

The first application module is called the **Comprehensive Measurement System** or CMS. It serves a wide range of data acquisition needs and illustrates techniques for voltage, time and counts measurements. Specifically, the CMS uses the CDI and ADC modules to:

(1) Determine the number of events in a succession of intervals.
(2) Determine the times between a succession of events.
(3) Determine the voltage from one or two sources either at a programmable rate or upon an external trigger.

The system receives up to six digital and two analog signals from experimental apparatus:

(1) Events to be counted.
(2) A start/stop for the counter.
(3) Events to determine the times between.
(4) A start/stop for the timer.
(5) A trigger for voltage measurement.
(6) A start/stop for voltage measurement.
(7) Voltage Input-A (to the ADC module).
(8) Voltage Input-B (to the ADC module).

The CMS circuit is built around three 8253 PIT's. PIT-A supports the counter, PIT-B the timer and PIT-C voltage measurement. Figure 9.1 shows the system in block form. The three 8253's are wired to the following CDI connections:

PIT Signal	CDI Connector
WR* (all PIT's)	CO_1 pin 3
RD* (all PIT's)	CO_2 pin 4
A_0 (all PIT's)	CO_6 pin 8
A_1 (all PIT's)	CO_7 pin 9
D7...D0 (all PIT's)	$d_7...d_0$ pins 28...21
PIT-A's CS*	CS_6* pin 5
PIT-B's CS*	CS_5* pin 6
PIT-C's CS*	CS_3* pin 7

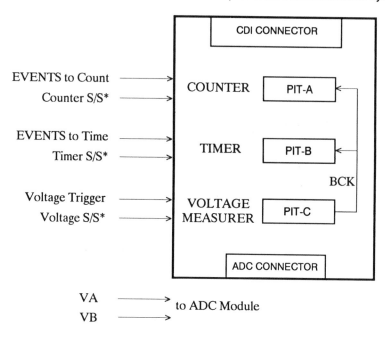

Figure 9.1. Block diagram of the Comprehensive Measurement System (CMS). BCK is a base clock generated by PIT-C and used by circuitry in all three measurement categories.

9.1.1 Base Clock (BCK)

All three measurement categories require a **Base ClocK** (BCK) which originates with PIT-C's CTR_0. Figure 9.2 gives the circuitry. A 1.0000 MHz digital oscillator is connected to CK_0, G_0 is wired high and BCK is OUT_0. The figure shows the PIT in a format which emphasizes the individual counters' clock and gate inputs and output. Except for CS_3*, the control and the data connections are not included.

To generate BCK, CTR_{C0} is programmed for mode 3 (as discussed in Section 6.2.2). As soon as a value is loaded, the output becomes a digital square wave whose frequency equals 1.0000 MHz divided by the loaded value. The following software sets up and starts the clock.

```
CALL PROGPIT(3,CW)
CALL LOADPIT(3,0,LOW,HIGH)
```

The subroutines are in the software shell in Section 7.7. The index 3 is used since CS_3* goes to PIT-C. CW is the mode Control Word. Recall from Section 6.4.2 that for CTR_0, mode 3 with 16-bit binary operation, CW = &H36. The value loaded in the storage register is 256*HIGH + LOW. Therefore, to make BCK 500 Hz, value must be 2000 which requires HIGH

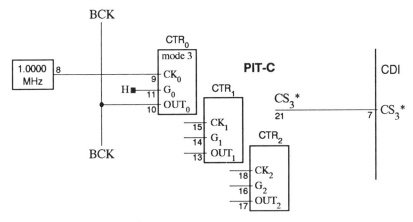

Figure 9.2. Circuitry which generates the base clock BCK. The PIT is represented in a new format which omits WR*, RD*, A_0, A_1 and D7...D0. The CDI connector is on the right.

= 7 and LOW = 208. At any time, BCK can be programmed for a different frequency by changing LOW and HIGH and executing

```
CALL LOADPIT(3,0,LOW,HIGH)
```

Possible load values range from 2 to 65,535. However, it's desirable for BCK to have integer frequencies which means values divide evenly into 1,000,000. Therefore, the practical range for BCK is 500 KHz (value = 2) down to 20 Hz (value = 50,000).

The following three sections present the design principles, circuitry, software and capabilities for the three measurement categories. Section 9.5 gives the complete CMS circuit. Section 9.6 illustrates ways to combine measurements. Section 9.7 describes construction. And Section 9.8 covers testing and troubleshooting.

9.2 Counter

The CMS counter determines the number of events in one or a succession of programmable length intervals. Example measurements are: (1) the number of detected radioactive decays each 100 ms for 10 seconds; (2) the number of flywheel rotations per second for 500 seconds; (3) the number of times a caged rat obtains water per hour for a day; and (4) the number of cycles of a high frequency electrical signal in a millisecond. In these and all other cases, information must be put in the from of digital signals with events represented as high-low transitions (as explained below). The analog electronics which converts sensor outputs to appropriate digital form is not covered in this book. However, references are in Appendix A.

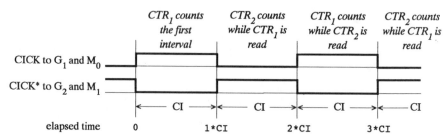

Figure 9.3. Timing diagram showing the gates to the PIT counters which alternate collecting data. CICK is the counting interval clock and CI is the counting interval.

When multiple counting intervals are used, it's often important that the end of one interval coincide with the beginning of the next. If CI is the length of the Counting Interval, this means the end of the nth interval is n*CI seconds after the beginning of the first. So, in essence, the system determines counts versus time.

The design approach is for PIT-A's CTR_1 and CTR_2 to acquire data. After initiation, CTR_1 counts the first interval, CTR_2 the second, CTR_1 the third, etc. The scheme is shown in Figure 9.3 and circuitry is in Figure 9.4. CICK (Counting Interval ClocK) and its inverse CICK* go to G_1 and G_2, respectively. The EVENTS signal goes to both CK_1 and CK_2. Since a PIT counter responds to high-low transitions at its clock input only when the gate is high, CTR_1 counts events when CICK is high and CTR_2 when CICK* is high. Since CTR_2 starts when CTR_1 stops and vice versa, no time is lost between intervals and elapsed time is simply the sum of interval lengths.

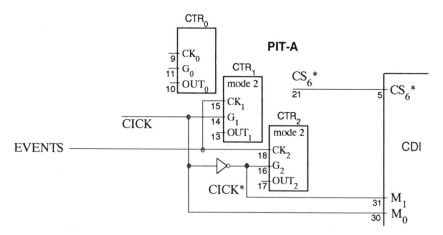

Figure 9.4. Circuitry which sets up the scheme described in Figure 9.3.

Recall from Chapter 6 that mode 2 is recommended for counting. Also, after every gate low-high, the first clock low-high-low copies the value loaded by software in the storage register to the count register. Subsequently, the count register decrements on every clock high-low. After a counting interval, the number of events equals the value loaded in the storage register minus the decremented value read from the count register plus 1. 8253 specifications require the clock signal to be high at least 230 ns before the high-low transition. Also, as discussed in Section 6.10, an error occurs if there are no events during an interval.

In the circuitry in Figure 9.4, CICK is also connected to CDI M_0 and CICK* to M_1. So, after initiation, software monitors M_0 until low is found and then reads CTR_1. Next, software monitors M_1 until low is found and then reads CTR_2. The process repeats for the desired number of intervals.

With this design framework, the remaining issues are: (1) how to set up and make the counting interval programmable; (2) how to use the CDI's S/ S* to initiate counting; and (3) how external hardware holds up and/or aborts counting.

9.2.1 Circuitry to Produce the Counting Interval

The counting interval is established by a combination of the base clock and PIT-A's CTR_0 (also programmed for mode 3). Figure 9.5 shows the additional circuitry. BCK goes to CK_0. G_0 is involved in initiation which is discussed next. OUT_0 originates CICK.

As can be seen in Figure 9.3, CICK's period is 2*CI (twice the length of the counting interval) which means its frequency is 1/(2*CI). CTR_{C0} and

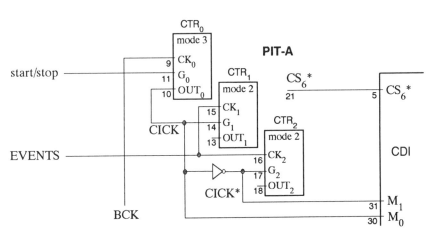

Figure 9.5. Circuitry which uses BCK to produce the counting interval clock CICK.

CTR_{A0} are loaded with values so CICK has the right frequency for the desired CI. The following table shows by example how to compute LOW and HIGH for both counters. The first column is the desired counting interval. The next two give the period and frequency of CICK. The next is the base clock frequency which should be an even divisor of 1,000,000. The next three columns give value, LOW and HIGH for CTR_{C0} so BCK = 1,000,000/value. The last three columns give value, LOW and HIGH for CTR_{A0} so CICK = BCK/value.

desired CI (sec)	CICK period (sec)	CICK freq (Hz)	BCK freq (Hz)	CTR-C0 value	CTR-C0 LOW	CTR-C0 HIGH	CTR-A0 value	CTR-A0 LOW	CTR-A0 HIGH
1.0	2.0	0.50	1000	1000	232	3	2000	208	7
1.0	2.0	0.50	100	10000	16	39	200	200	0
0.001	0.002	500.00	1000	1000	232	3	2	2	0
0.010	0.020	50.0	1000	1000	232	3	20	20	0
0.025	0.050	20.0	1000	1000	232	3	50	50	0
0.10	0.200	5.0	1000	1000	232	3	200	200	0
100.0	200.0	0.005	100	10000	16	39	20000	32	78
1250.0	2500.0	0.0004	20	50000	80	195	50000	80	195
60.0	120.0	0.00833	500	2000	208	7	60000	96	234
1/60	0.0333	30.0	500000	2	2	0	16667	27	65

The first eight rows of the table cover "even" cases. The last two are for CI = 60 seconds and $1/60^{th}$ of a second. CICK for 60 seconds rounds off to .00833 Hz. However, if 500 Hz is picked for BCK, then 500/.00833 is exactly 60,000. For the $1/60^{th}$ case, CICK is exactly 30 Hz and no multiple of 30 divides evenly into 1,000,000. The least error occurs if BCK is as large as possible. For BCK = 500,000 KHz, value = 500000/30 = 16,666.6 which is extremely close to the nearest integer 16,667. If BCK were smaller, the round-off would be proportionally larger.

Longer intervals are possible with an additional PIT (which is not available in the CMS but could be in other applications modules).

9.2.2 Counter Initiation

It's desirable to be able to initiate counting sequences with either software or hardware. It's also desirable for hardware to be able to abort a sequence. Figure 9.6 gives circuitry which accomplishes this. S-S/S* (Software Start/Stop) from the CDI is ANDed with EX-S/S* (EXternal Start/Stop) from the experimental apparatus. The result C-S/S* (Counter Start/Stop) goes to G_0. When C-S/S* goes high, OUT_0 becomes the clock shown in Figure 9.3.

Suppose EX-S/S* is wired high. S-S/S* is ordinarily low (the first thing application programs do is CALL SSTOP). When S-S/S* goes high (via CALL START), C-S/S* goes high and data collection begins. On the other hand, if EX-S/S* is low, C-S/S* stays low when S-S/S* goes high. Counting

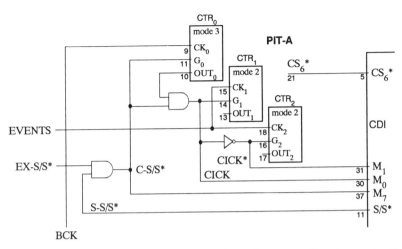

Figure 9.6. Circuitry which combines the software and hardware start/stops to produce the counter start/stop.

is held up until EX-S/S* goes high and thereby external hardware controls initiation.

There is a fundamental problem with this scheme. In all modes, a PIT counter's output is high when the gate is low (see Figure 6.7). This means OUT_0 is high before data acquisition starts (and CTR_1 prematurely counts events). The problem is solved by ANDing OUT_0 with C-S/S* to create CICK (G_1) and CICK* (G_2). This keeps G_1 low until C-S/S* starts the first interval.

C-S/S* also goes to the CDI's M_7 input. Recall from Chapter 7 that control-in bit 7 is the result of ORing CDI inputs M_7, M_8 and M_9. If the latter two are wired low, then CI_7 = C-S/S*. Therefore, software can respond if the experiment takes EX-S/S* low during data collection.

There is another problem with the circuit in Figure 9.6. As explained in Section 6.6.2, after the gate to a mode 3 PIT goes high, the first output cycle is delayed until the signal at the clock input goes low-high-low. This means the first OUT_0 cycle doesn't start when C-S/S* (G_0) goes high, but later after BCK has gone low-high-low. However, because OUT_0 is already high, G_1 (CICK) goes high with C-S/S* and prematurely starts the first count interval. Additional circuitry is needed to prevent this.

The problem does not affect G_1 after the first interval and never affects G_2. In those cases, G_0 hasn't just gone high and the copy process is instantaneous and automatic (as explained in Section 6.6.2). Also, the delay is not significant if the counting interval is long.

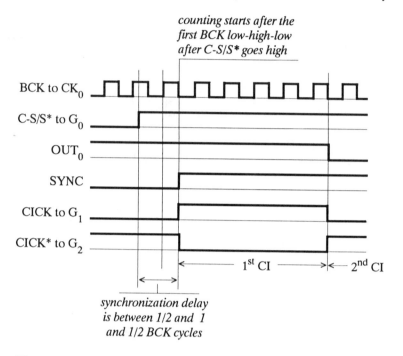

counting starts after the
first BCK low-high-low
after C-S/S* goes high

synchronization delay
is between 1/2 and 1
and 1/2 BCK cycles

Figure 9.7. CICK and the first counting interval must be delayed after C-S/S* goes high until CTR_0's storage register is copied to the count register. For the case shown, the value loaded in CTR_0 is 10.

Figure 9.7 shows the situation in an extension of the timing diagram in Figure 9.3. C-S/S* goes high at a random point in a BCK cycle. The new signal SYNC starts low and goes high at the end of BCK's low-high-low. So, if OUT_0 is ANDed with SYNC (rather than C-S/S*), counting would initiate properly. The job is to devise circuitry which generates SYNC.

The complete counter circuit is in Figure 9.8. The two flip-flops produce SYNC. As long as C-S/S* is low, both FF-A and B are clear and SYNC is low. When C-S/S* goes high, nothing happens. But the next time BCK goes high, FF-A is set. With Q_A high, FF-B is free to respond to its clock/data inputs. When BCK goes back low, BCK* goes high setting FF-B and making SYNC high, as shown in Figure 9.7. SYNC stays high until C-S/S* returns low.

Although not required, SYNC is ANDed with the EVENTS signal so neither CTR_1 nor CTR_2 count until the beginning of the first interval when G_1 is high and G_2 low.

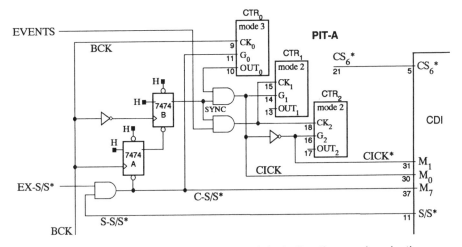

Figure 9.8. The complete counter circuit including the synchronization flip-flops.

9.2.3 Counter Software

The counter circuit is initialized and controlled by the test/demonstration program developed in this section. The starting point is the software shell in Chapter 7.

CMS.BAS

Since the CMS module only uses PIT-A (CS_6*), PIT-B (CS_5*) and PIT-C (CS_3*), constants for the other PIT's (0, 1, 2 and 4) are deleted from the shell. The memory constants and subroutines are not used and are also deleted. The modified shell is saved as CMS.BAS. The remaining subroutines are: SSTOP, START, PROGPIT(I,CW), LOADPIT(I,J,LOW,HIGH), READPIT (I,J,LOW,HIGH).

Constants and Arrays

Four PIT counters must be programmed and loaded. Recall from Chapter 7 that each has a unique index I,J. I equals the number of the CDI chip select connected to the PIT and J is the counter. The following table gives each counter's mode, index, control word, LOW and HIGH load bytes and function.

counter	mode	index	CW	LOW	HIGH	function
C0	3	3,0	&H36	X	X	generate BCK
A0	3	6,0	&H36	X	X	generate CKCI
A1	2	6,1	&H74	&HFF	&H7F	count events
A2	2	6,2	&HB4	&HFF	&H7F	count events

"X" in the table means the value is determined later. As discussed in Section 6.8.2, mode 2 and a load value of 7FFF hex are recommended for data collection counters.

It's efficient to set up arrays for the CW's, LOW's and HIGH's. Also, data are stored in the array COUNT(I) which is dimensioned for the maximum anticipated number of counting intervals. The following declarations are added to CMS.BAS.

```
DIM LOW(7,3),HIGH(7,3),CW(7,3)
DIM COUNT(1000)
```

The following constants are assigned values from the table.

```
CW(3,0) = &H36
CW(6,0) = &H36
CW(6,1) = &H74 : LOW(6,1)= &HFF : HIGH(6,1)= &H7F
CW(6,2) = &HB4 : LOW(6,2)= &HFF : HIGH(6,2)= &H7F
```

Initialization

All CMS application programs begin with

```
MAIN: CALL SSTOP
```

which establishes a dormant state for the hardware.

The next task is to determine the counting interval and assign values to LOW(3,0), HIGH(3,0), LOW(6,0) and HIGH(6,0). There are three approaches. If CI is known when the program is written, values are just assigned. If CI is entered during execution, the program must compute CICK, BCK, load values for both counters, and the LOW's and HIGH's. (The task is tricky because of round-off errors and should be thoroughly tested.) The case of selecting from a limited sets of CI's is illustrated here.

```
CLS
PRINT "CI Values in Seconds: 1.0, 0.1, 0.01 "
INPUT "Enter Selection "; CI!
```

The "!" appended to CI overrides the DEFINT A-Z declaration and makes the variable real. LOW and HIGH for these CI's are in the table in the last section. BCK equals 1000 Hz for each. The following program statements assign values to the constants.

```
LOW(3,0)  = 232 : HIGH(3,0) = 3
IF CI! = 1.0 THEN LOW(6,0) = 208 : HIGH(6,0) = 7
IF CI! = 0.1 THEN LOW(6,0) = 200 : HIGH(6,0) = 0
IF CI! = 0.01 THEN LOW(6,0) = 20 : HIGH(6,0) = 0
```

Another parameter is the Number of counting INTervals (NINT). The easiest data collection scheme is a loop which waits for and reads values first from CTR_1 and then from CTR_2. With two intervals per loop, NINT should be even.

```
INPUT "Enter an Even Number of Intervals "; NIN
```

The next operations program and load the counters.

```
CALL PROGPIT(3,CW(3,0))
CALL LOADPIT(3,0,LOW(3,0),HIGH(3,0))
FOR J = 0 TO 2
  CALL PROGPIT(6,CW(6,J))
  CALL LOADPIT(6,J,LOW(6,J),HIGH(6,J))
NEXT J
```

This completes initialization.

Data Collection Loop

After calling start, software must wait until EX-S/S* goes high (if necessary) and then wait during the synchronization delay. So, the first operation is wait while M_0 ($=G_1$) is low (as shown in Figure 9.7.)

```
        CALL START
WSYNC: IF (INP(CP) AND &H01) = 0 THEN GOTO WSYNC
```

The WSYNC (**W**ait for **SYNC**hronization) statement repeats as long as M_0 is low. Specifically, INP(CP) obtains from the control-in port X X X X X X X M0 which after ANDing with 00000001 equals 0 0 0 0 0 0 0 M0 . If M_0 is high, the result is 1; and if M_0 is low, the result is 0.

> *In slow computers when EX-S/S* is not used, the SYNC delay might be over (M_0 is high) before WSYNC is executed. This causes no problem since the next loop is immediately entered. On the other hand, if program logic were to wait until M_0 is high, an infinite loop would occur.*

Next, the program enters a STEP 2 loop which repeats NINT/2 times. The first operation WCTR1 (**W**ait for **CTR**$_1$ data) repeats while M_0 is high. Then CTR_1 is read. Next, CI_7 is tested for an external abort. After that, WCTR2 loops while M_1 ($=G_2$) is high after which CTR_2 is read. External abort is again evaluated. COUNT(I) is computed as discussed in Chapter 6.

```
        FOR I = 1 TO NINT STEP 2
WCTR1:  IF (INP(CP) AND &H01) = 1 THEN GOTO WCTR1
        CALL READPIT(6,1,DLOW,DHIGH)
        COUNT(I) = &H7FFF - (256*DHIGH + DLOW) + 1
        IF (INP(CP) AND &H80) = 0 THEN GOTO ABORT
WCTR2:  IF (INP(CP) AND &H02) = 2 THEN GOTO WCTR2
        CALL READPIT(6,2,DLOW,DHIGH)
        COUNT(I+1) = &H7FFF - (256*DHIGH + DLOW) + 1
        IF (INP(CP) AND &H80) = 0 THEN GOTO ABORT
        NEXT I
        CALL SSTOP
```

The mask for M_1 is 00000010 or 2. ABORT labels the repeat option given later.

The shortest counting interval is determined by how fast the computer and language execute the statements between WCTR1 and WCTR2. This time is estimated later. On a 12.5 MHz PC-AT compatible running Quick-BASIC, the shortest "safe" interval for the program above is 1.5 ms. However, for the program converted by QuickBASIC to a stand-alone .EXE form, the shortest CI is 200 μsec.

Execution may be speeded up by saving DLOW and DHIGH in separate arrays (in the READPIT statements) and deferring computations until after data acquisition. Also, the check for external abort might be removed.

Display Results and Repeat

After acquisition, count values and elapsed times (to the end of each interval) are displayed as follows.

```
CLS
PRINT "Results from CI = ";CI!;" seconds"
PRINT "# of Events During CI,Elapsed Time in Sec"
FOR I = 1 TO NINT
    PRINT COUNT(I);" , ";I * CI!,
NEXT I
ABORT: INPUT "Enter R to Repeat, Else Program Ends ";A$
IF A$ = "R" OR A$ = "r" THEN GOTO MAIN
END
```

Complete Test/Demonstration Program

The bits and pieces above are entered into CMS.BAS to produce the following program. Because the counter does not use PIT-B (CS_5*) or read any decives, constants for both are deleted.

```
'       Program: TEST COUNTER

DEFINT A-Z
DIM SELPP(7),PROGP(7)
DIM SELPC(7,3),LOADPC(7,3),READPC(7,3)
DIM LOW(7,3), HIGH(7,3),CW(7,3)
DIM COUNT(1000)
COMMON SHARED CP,DP,NORM,STRT,SSTP
COMMON SHARED SELPP(),PROGP(),SELPC()
COMMON SHARED LOADPC(),READPC()
CP = &H300 : DP = &H301
NORM = &HFF
STRT = &HC6 : SSTP = &HCE
CW(3,0) = &H36
SELPP(3) = &HDF
PROGP(3) = &HDD
```

```
        SELPC(3,0) = &H1F : SELPC(3,1) = &H5F
        SELPC(3,2) = &H9F
        LOADPC(3,0) = &H1D : LOADPC(3,1) = &H5D
        LOADPC(3,2) = &H9D
        READPC(3,0) = &H1B : READPC(3,1) = &H5B
        READPC(3,2) = &H9B
        CW(6,0) = &H36
        CW(6,1) = &H74 : LOW(6,1) = &HFF
        HIGH(6,1) = &H7F
        CW(6,2) = &HB4 : LOW(6,2) = &HFF
        HIGH(6,2) = &H7F
        SELPP(6) = &HF7
        PROGP(6) = &HF5
        SELPC(6,0) = &H37 : SELPC(6,1) = &H77
        SELPC(6,2) = &HB7
        LOADPC(6,0) = &H35 : LOADPC(6,1) = &H75
        LOADPC(6,2) = &HB5
        READPC(6,0) = &H33 : READPC(6,1) = &H73
        READPC(6,2) = &HB3

MAIN:   CALL SSTOP
        CLS
        PRINT "CI Values in Seconds: 1, 0.1, 0.01 "
        INPUT "Enter Selection "; CI!
        LOW(3,0) = 232 : HIGH(3,0) = 3
        IF CI! = 1 THEN LOW(6,0) = 208 : HIGH(6,0) = 7
        IF CI! = .1 THEN LOW(6,0) = 200 : HIGH(6,0) = 0
        IF CI! = .01 THEN LOW(6,0) = 20 : HIGH(6,0) = 0
        INPUT "Enter an Even Number of Intervals "; NINT

        CALL PROGPIT(3,CW(3,0))
        CALL LOADPIT(3,0,LOW(3,0),HIGH(3,0))
        FOR J = 0 TO 2
          CALL PROGPIT(6,CW(6,J))
          CALL LOADPIT(6,J,LOW(6,J),HIGH(6,J))
        NEXT J

        CALL START
WSYNC:  IF (INP(CP) AND &H01) = 0 THEN GOTO WSYNC

        FOR I = 1 TO NINT STEP 2
WCTR1:    IF (INP(CP) AND &H01) = 1 THEN GOTO WCTR1
          CALL READPIT(6,1,DLOW,DHIGH)
          COUNT(I) = &H7FFF - (256*DHIGH + DLOW) + 1
          IF (INP(CP) AND &H80) = 0 THEN GOTO ABORT
WCTR2:    IF (INP(CP) AND &H02) = 2 THEN GOTO WCTR2
          CALL READPIT(6,2,DLOW,DHIGH)
          COUNT(I+1) = &H7FFF - (256*DHIGH + DLOW) + 1
          IF (INP(CP) AND &H80) = 0 THEN GOTO ABORT
        NEXT I

        CALL SSTOP
        CLS
```

```
      PRINT "Results from CI = ";CI!;" seconds"
      PRINT "# of Events During CI,Elapsed Time in Sec"
      FOR I = 1 TO NINT
        PRINT COUNT(I);" , ";I * CI!,
      NEXT I
ABORT: INPUT "Enter R to Repeat, Else Program Ends ";A$
      IF A$ = "R" OR A$ = "r" THEN GOTO MAIN
      END
```

The subroutines SSTOP, START, PROGPIT, LOADPIT and READPIT
follow but are not reproduced here.

Estimation of the Minimum Counting Interval

A variation of the test/demonstration program can be used to roughly determine the minimum counting interval. Remove the six program lines starting
two lines after MAIN:. Add the following statements:

```
      PRINT "Enter LOW and HIGH for BCK "
      INPUT LOW(3,0),HIGH(3,0)
      PRINT "Enter LOW and HIGH for CICK "
      INPUT LOW(6,0),HIGH(6,0)
```

Use a 100 KHz digital clock for the EVENTS signal. Run the program.
Enter values from the table in Section in 9.2.1 so CI is .01 second. Select 10
counting intervals. 1000 counts should occur in each interval. Then, systematically reduce CI (CICK's frequency) until the expected number of counts is
not found. This happens when CI is so short that the next interval starts
before the previous data is read.

> *In addition to estimating the minimum CI, the modified program is a good
> exercise in computing load values for various CI's.*

9.2.4 Counter Capabilities and Limitations

The counter's capabilities and limitations are:
(1) The shortest counting interval (CI) depends on what operations are
     carried out in the data collection loop and on the speed of the computer
     and language. A typical value for PC-AT's using QuickBASIC is 1 ms.
(2) The longest CI is around 1250 seconds.
(3) Accuracy for an interval is ±1 count (as discussed in Section 6.3).
(4) The maximum number of counting intervals is limited by the memory
     available for data storage.
(5) The maximum number of events in an interval is 32K (since 7FFF hex is
     the recommended value to count down from). Capacity could be ex-

tended to 64K by loading FFFF hex and using a double precision integer variables to compute the number of counts:

```
DDLOW# = CDBL(DLOW)
DDHIGH# = CDBL(DHIGH)
COUNT# = 65535 - 256*DDHIGH# + DDLOW# + 1
```

An appended "#" makes QuickBASIC variables double precision. The function CDBL converts the single precision integers DLOW and DHIGH (returned from READPIT) to double precision. Hex numbers default to single precision. Therefore, 65535 must be used in place of &HFFFF.

(6) Finally, the EVENTS signal's maximum frequency is around 2 MHz (8253 specifications, say, the signal at the clock input must be high 230 ns before going low).

In addition to these capabilities and limitations, the following cautions must be observed:

(1) If no events occur in a counting interval, 7FFF is not copied from the storage to the count register. The subsequent READPIT gets whatever value was left by previous operations. Therefore, action must be taken if there is the remotest possibility of no events. Solutions were suggested in Section 6.10.

(2) There is no warning if more than 32K counts occur during an interval. Since PIT counters decrement with every clock high-low and roll-over from 0000 to FFFF hex, computations in this case may or may not cause "overflows."

(3) Finally, if CI is too short and the next gate high interval starts before LOW and HIGH are read, the values obtained are erroneous. The possibility is dealt with by knowing the shortest interval and being cautious.

9.2.5 Example Measurement

The test/demonstration program determines the number of events in a set of CI's and displays the counts and elapsed times. The purpose of this section is to illustrate how software creates a particular instrument. Suppose it's necessary to continually know (to within 0.1%) the frequency of a digital signal originating from some process. The frequency varies between 500 Hz and 500 KHz.

An approach is to set a counting interval, determine the number of cycles (of the signal) during one interval and then compute the frequency from #cycles/time = counts/CI. The first issue is what interval to use. For instance, if CI is 1.0 second and the frequency 500 KHz, 500,000 events occur which far exceeds counter capacity. On the other hand, if CI is 0.002 second and the frequency 500 Hz, there is a chance for no events during CI. Therefore, different CI's must be used for different frequencies.

To achieve an accuracy of 0.1%, there must be at least 1000 counts in an interval. For frequencies between 500 Hz and 500 KHz, CI's of 0.05, 0.50 and 5.00 seconds accomplish the task as shown in the following table. (Other cases also work.)

CI (seconds)	if the frequency is	counts
0.05	500. KHz	25000
0.05	20. KHz	1000
0.05	500. Hz	25
0.50	20. KHz	10000
0.50	2. KHz	1000
5.00	2. KHz	10000
5.00	500. Hz	2500

CI = 0.05 second produces 25000 counts at the maximum frequency and 1000 counts at 20 KHz. CI = 0.50 produces 10000 counts at 20 KHz and 1000 counts at 2 KHz. And CI = 5.00 produces 10000 counts at 2 KHz and 2500 at the minimum frequency.

The scheme is to set CI initially to 0.05 second and determine the Number of **CYCLES** (NCYCLES) in the interval. The value will be between 25,000 and 25. If NCYCLES is greater than 1000, sufficient accuracy exists and the frequency is computed and displayed. If NCYCLES is between 100 and 1000, CI is programmed for 0.5 second and a new NCYCLES acquired. It should be greater than 1000 and is displayed. And if the initial NCYCLES is less than 100, CI is programmed for 5.00 seconds and NCYCLES determined. It should be greater than 1000 and the result is displayed. (It's assumed the frequency does not change appreciably during the time required for this sequence.)

The following table gives the CTR_{A0} (CICK) load values for the three cases. BCK is 1000 Hz and need not be changed.

desired CI (sec)	CICK period (sec)	freq (Hz)	BCK freq (Hz)	CTR-C0 value	LOW	HIGH	CTR-A0 value	LOW	HIGH
0.050	0.10	10.0	1000	1000	232	3	100	100	0
0.50	1.00	1.0	1000	1000	232	3	1000	232	3
5.00	10.00	0.10	1000	1000	232	3	10000	16	39

The following program continually displays frequency and automatically adjusts to changes.

```
PROGRAM: Measure Frequency

DEFINT A-Z

DIM SELPP(7),PROGP(7)
DIM SELPC(7,3),LOADPC(7,3),READPC(7,3)
DIM LOW(7,3),HIGH(7,3),CW(7,3)
```

```
      COMMON SHARED CP,DP,NORM,STRT,SSTP
      COMMON SHARED SELPP(),PROGP()
      COMMON SHARED SELPC(),LOADPC(),READPC()

      CP = &H300 : DP = &H301
      NORM = &HFF
      STRT = &HC6 : SSTP = &HCE

      CW(3,0) = &H36
      SELPP(3) = &HDF
      PROGP(3) = &HDD
      SELPC(3,0) = &H1F
      LOADPC(3,0) = &H1D
      READPC(3,0) = &H1B

      CW(6,0) = &H36
      CW(6,1) = &H74: LOW(6,1) = &HFF: HIGH(6,1) = &H7F
      CW(6,2) = &HB4: LOW(6,2) = &HFF: HIGH(6,2) = &H7F
      SELPP(6) = &HF7
      PROGP(6) = &HF5
      SELPC(6,0) = &H37: SELPC(6,1) = &H77
      SELPC(6,2) = &HB7
      LOADPC(6,0) = &H35: LOADPC(6,1) = &H75
      LOADPC(6,2) = &HB5
      READPC(6,0) = &H33: READPC(6,1) = &H73
      READPC(6,2) = &HB3

MAIN: CALL SSTOP
      CLS
      CALL PROGPIT(3,CW(3,0))
      CALL LOADPIT(3,0,232,3)
      FOR J = 1 TO 2
        CALL PROGPIT(6,CW(6,J))
        CALL LOADPIT(6,J,LOW(6,J),HIGH(6,J))
      NEXT J
      CALL PROGPIT(6,CW(6,0))

REPT1:CI! = .05
      CALL LOADPIT(6,0,100,0)
REPT2:CALL START
WSYNC:IF (INP(CP) AND &H01) = 0 THEN GOTO WSYNC
WCTR1:IF (INP(CP) AND &H01) = 1 THEN GOTO WCTR1
      CALL SSTOP
      CALL READPIT(6,1,DLOW,DHIGH)
      NCYCLES = &H7FFF - (256 * DHIGH + DLOW) + 1

      IF INKEY$ <> "" THEN GOTO FINI

      IF NCYCLES >= 1000 THEN GOTO DFREQ

      IF NCYCLES >= 100 THEN
       CALL LOADPIT(6,0,232,3)
       CI! = .5
       GOTO REPT2
      END IF
```

```
        IF NCYCLES >= 10 THEN
          CALL LOADPIT(6,0,16,39)
          CI! = 5!
          GOTO REPT2
        END IF

        PRINT "Count Less than 200 Hz - Program Aborted "
        GOTO FINI
DFREQ: LOCATE 10,10
        FREQ! = NCYCLES/CI!
        PRINT "Frequency in Hz ";
        IF CI! = .05 THEN PRINT USING "#######"; FREQ!
        IF CI! = .5 THEN PRINT USING "######."; FREQ!
        IF CI! = 5 THEN PRINT USING "#####.#"; FREQ!
        GOTO REPT1

FINI:  END
```

The subroutines SSTOP, START, PROGPIT, LOADPIT and READPIT follow but are not reproduced here.

After a frequency is displayed, the program resets CI to 0.05 second and repeats. If this were not done, an increasing signal frequency might produce more than 32K counts in one of the longer intervals. The INKEY statement allows termination of repeated measurement.

The case illustrated here is not sophisticated and is intended only to suggest possibilities. More elaborate reprogramming and averaging schemes are possible. Even things such as how fast the frequency is changing could be evaluated by acquiring data for a set of intervals and observing differences. Finally, if adjusting software is not adequate to meet a particular measurement need, a specialized application module could be designed using principles illustrated here.

9.3 Timer

The CMS timer determines the period between events. When more than two events are involved, each (except the first and last) marks the end of one timing period and the beginning of the next. This makes the elapsed time from the first to the nth event the sum of the preceding periods. Example measurements are: (1) the period of each flywheel rotation for 100 rotations (from which angular velocity vs. time is computed); (2) the net time required for 1000 radioactive decays; (3) the average time between activations of an error signal over a day; and (4) the time light is blocked from a phototransistor. In these and all other cases, information must be put in the form of digital signals with events represented as low-high transitions.

9.3.1 Timer Hardware

The design approach is for PIT-B's CTR_1 and CTR_2 to alternate counting the number of cycles of a known frequency clock between events. Time then equals counts multiplied by the clock's period. The Timing ClocK (TCK) is established by the base clock (CTR_{C0}) and by CTR_{B0}. Figure 9.9 shows the scheme. EVENTS occur on low-to-high transitions and are numbered 0, 1, 2, 3, etc. Event 0 takes G_1 high during which CTR_1 counts TCK cycles. Event 1 takes G_1 low and G_2 high. Then while CTR_2 is counting, CTR_1 is read and T_1 computed. Next, event 2 takes G_2 low and G_1 high. While G_1 is again counting, G_2 is read and T_2 computed. And so on.

Figure 9.10 gives the timer circuit. The EVENTS signal goes to the clock input of a toggling flip-flop. The Q and Q* outputs go to G_1 and G_2, respectively. Therefore, each event (low-high transition) causes Q and Q* to reverse and the counters to alternate taking data. CTR_1 and CTR_2 are programmed for mode 2 and loaded with 7FFF hex.

G_1 is connected to the CDI's M_2. By monitoring the line, software knows when to read CTR_1. Similarly, G_2 is connected to M_3 so software knows when CTR_2 data is ready.

BCK goes to CK_0. OUT_0 originates TCK which goes to CK_1 and CK_2 (enabling CTR_1 and CTR_2 to count cycles when their gates are high). CTR_{B0} is programmed for mode 3, as is CTR_{C0}. Values loaded into both counters determine TCK's frequency as discussed in the next section.

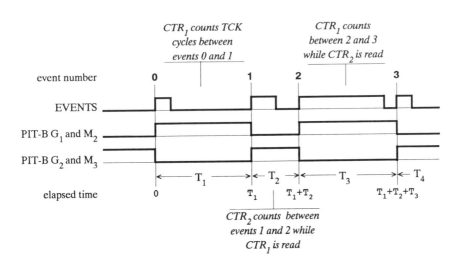

Figure 9.9. Timing diagram showing the EVENTS signal and corresponding gates to the PIT counters which alternate collecting data. Events occur on low-to-high transitions. TCK is the timing clock.

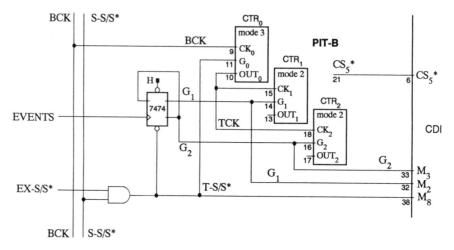

Figure 9.10. The timer circuit. BCK originates with PIT-C CTR$_0$. S-S/S*
originates with the CDI.

Initiation works as follows. Software Start/Stop (S-S/S*) and **EX**ternal
hardware Start/Stop (EX-S/S*) are ANDed. The result T-S/S* (Timer Start/
Stop) goes to the flip-flop's CLEAR* and to G$_0$. While S-S/S* and therefore
T-S/S* are low (indicating stop), Q (G$_1$) is low and TCK is not produced. If
EX-S/S* is high when S-S/S* goes high, TCK becomes available to both
counters. However, nothing happens to the flip-flop and G$_1$ stays low. The
next event makes G$_1$ high and the pattern in Figure 9.9 begins. On the other
hand, if ES-S/S* is low when S-S/S* goes high, initiation is held up by external hardware. Finally, T-S/S* is also connected to CDI's M$_8$. If M$_7$ and M$_9$
are low, then software via CI$_7$ can monitor T-S/S* and EX-S/S* during data
acquisition.

9.3.2 TCK Frequencies

TCK's frequency is set in anticipation of the times to be measured. For
instance, a 5000 Hz TCK undergoes 5000 cycles in 1 second, 5 cycles in 1 ms
and 500,000 cycles in 100 seconds. This section illustrates how to select appropriate TCK values.

Recall from Section 6.3 that the number of counts is accurate to ±1. So
100 TCK cycles per timing period are required for 1% results. The following
table shows how to set up three time ranges which start at 1 ms and go up to
10,000 seconds. The first column is the range. The next two give the TCK
period and frequency necessary to get 100 counts for the shortest time in the
range. (The longest time in each range was selected so less than the maximum 32K TCK counts occur.) The fourth column is a reasonable base clock

frequency. The remaining six columns give value, LOW and HIGH for both counters.

time range to be measured	TCK for at least 100 cycles period (ms)	freq (Hz)	BCK freq (Hz)	CTR-C0 value	LOW	HIGH	CTR-B0 value	LOW	HIGH
1 to 200 ms	0.01 ms	100. KHz	500000	2	2	0	5	5	0
.2 to 50 sec	2.0 ms	500. Hz	500000	2	2	0	1000	232	3
50 to 10,000	0.5 sec	2. Hz	10000	100	100	0	5000	136	19

Obviously, other ranges are possible and accuracy can be increased or decreased as desired.

9.3.3 Timer Software

The following test/demonstration program introduces software which operates the timer circuit. The starting point is CMS.BAS (introduced in Section 9.2.3).

Constants and Arrays

Four PIT's must be programmed and loaded almost exactly the same as for the counter. The following table summarizes the task. Recall that PIT-B is connected to the CDI's CS_5*.

counter	mode	index	CW	LOW	HIGH	function
C0	3	3,0	&H36	X	X	generate BCK
B0	3	5,0	&H36	X	X	generate TCK
B1	2	5,1	&H74	&HFF	&H7F	count TCK cycles
B2	2	5,2	&HB4	&HFF	&H7F	count TCK cycles

"X" means the value is determined later.
   Arrays not in CMS.BAS are:

```
DIM LOW(7,3),HIGH(7,3),CW(7,3)
TI(1000)
```

where TI(1000) receives Time Interval data.
   The following constants are assigned values from the table:

```
CW(3,0) = &H36
CW(5,0) = &H36
CW(5,1) = &H74 : LOW(5,1)= &HFF : HIGH(5,1)= &H7F
CW(5,2) = &HB4 : LOW(5,2)= &HFF : HIGH(5,2)= &H7F
```

Initialization

All CMS programs begin with:

```
CALL SSTOP
```

The next task is to set the timing clock's frequency and assign values to LOW(3,0), HIGH(3,0), LOW(5,0) and HIGH(5,0). The following software implements selection of one of the three ranges introduced earlier.

```
CLS
PRINT "Available Time Ranges"
PRINT "1: 1 ms to 200 ms"
PRINT "2: 200 ms to 50 sec"
PRINT "3: 50 sec to 10,000 sec"
INPUT "Enter Number of Desired Range "; RANGE

IF RANGE = 1 THEN
  TCK! = 100000
  LOW(3,0) = 2: HIGH(3,0) = 0
  LOW(5,0) = 5: HIGH(5,0) = 0
END IF

IF RANGE = 2 THEN
  TCK! = 500
  LOW(3,0) = 2: HIGH(3,0) = 0
  LOW(5,0) = 232: HIGH(5,0) = 3
END IF

IF RANGE = 3 THEN
  TCK! = 2
  LOW(3,0) = 100: HIGH(3,0) = 0
  LOW(5,0) = 136: HIGH(5,0) = 19
END IF

PTCK! = 1/TCK!
```

TCK! is the frequency of the timing clock and PTCK! is its Period.

Complete Test/Demonstration Program

The remaining software is nearly identical to that for the counter. Exceptions are:

(1) The timer circuit has no synchronization delay. But still the operation just before the data collection loop is to wait while M_2 (G_1) is low. This is necessary in case EX-S/S* holds up data collection.

(2) NEVT/2 (Number of EVenTs) rather than NINT/2 is the number of acquisition loops.

(3) WFEVT (Wait for the First EVenT) loops while M_2 is low. The mask is 00000100 (4). WCTR1 (Wait for data from CTR_1) loops while M_2 is high (the AND result is not equal to 0). WCTR2 (Wait for data from CTR_2) loops while M_3 is high (the mask is 00001000 or 8).

(4) The number of cycles in a timing interval is computed from DLOW and DHIGH (obtained by READPIT) and saved in the array TI(I).

(5) When results are displayed, the **INTERVAL** and Elapsed **TIME**s are computed:

```
          INTERVAL! = TI(I) * PTCK!
          ETIME! = ETIME! + INTERVAL!
```

ETIME starts at 0.

(6) Constants for PIT-A (CS_6*) are deleted as are the read device subroutine and constants.

The complete program is:

```
'   PROGRAM:  Test Timer

    DEFINT A-Z
    DIM SELPP(7),PROGP(7)
    DIM SELPC(7,3),LOADPC(7,3),READPC(7,3)
    DIM REDD(7)
    DIM LOW(7,3),HIGH(7,3),CW(7,3)
    DIM TI(1000)
    COMMON SHARED CP,DP,NORM,STRT,SSTP
    COMMON SHARED SELPP(),PROGP(),SELPC()
    COMMON SHARED LOADPC(),READPC()

    CP = &H300 : DP = &H301
    NORM = &HFF
    STRT = &HC6 : SSTP = &HCE

    CW(3,0) = &H36
    SELPP(3) = &HDF
    PROGP(3) = &HDD
    SELPC(3,0) = &H1F : SELPC(3,1) = &H5F
    SELPC(3,2) = &H9F
    LOADPC(3,0) = &H1D : LOADPC(3,1) = &H5D
    LOADPC(3,2) = &H9D
    READPC(3,0) = &H1B : READPC(3,1) = &H5B
    READPC(3,2) = &H9B

    CW(5,0) = &H36
    CW(5,1) = &H74 : LOW(5,1) = &HFF
    HIGH(5,1) = &H7F
    CW(5,2) = &HB4 : LOW(5,2) = &HFF
    HIGH(5,2) = &H7F
    SELPP(5) = &HEF
    PROGP(5) = &HED
    SELPC(5,0) = &H2F : SELPC(5,1) = &H6F
    SELPC(5,2) = &HAF
    LOADPC(5,0) = &H2D : LOADPC(5,1) = &H6D
    LOADPC(5,2) = &HAD
    READPC(5,0) = &H2B : READPC(5,1) = &H6B
    READPC(5,2) = &HAB

MAIN: CALL SSTOP
      CLS
      PRINT "Available Time Ranges"
      PRINT "Range 1: 1 ms to 200 ms"
      PRINT "Range 2: 200 ms to 50 sec"
      PRINT "Range 3: 50 sec to 10,000 sec"
      INPUT "Enter Number of Desired Range "; RANGE
```

```
        IF RANGE = 1 THEN
          TCK! = 100000
          LOW(3,0) = 2 : HIGH(3,0) = 0
          LOW(5,0) = 5 : HIGH(5,0) = 0
        END IF
        IF RANGE = 2 THEN
          TCK! = 500
          LOW(3,0) = 2 : HIGH(3,0) = 0
          LOW(5,0) = 232 : HIGH(5,0) = 3
        END IF
        IF RANGE = 3 THEN
          TCK! = 2
          LOW(3,0) = 100 : HIGH(3,0) = 0
          LOW(5,0) = 136 : HIGH(5,0) = 19
        END IF
        PTCK! = 1/TCK!

        INPUT "Enter an Even Number of Events "; NEVT
        CALL PROGPIT(3,CW(3,0))
        CALL LOADPIT(3,0,LOW(3,0),HIGH(3,0))
        FOR J = 0 TO 2
          CALL PROGPIT(5,CW(5,J))
          CALL LOADPIT(5,J,LOW(5,J),HIGH(5,J))
        NEXT J

        CALL START
WFEVT: IF (INP(CP) AND &H04) = 0 THEN GOTO WFEVT

        FOR I = 1 TO NEVT STEP 2
WCTR1:    IF (INP(CP) AND &H04) = 4 THEN GOTO WCTR1
          CALL READPIT(5,1,DLOW,DHIGH)
          TI(I) = &H7FFF - (256 * DHIGH + DLOW) + 1
WCTR2:    IF (INP(CP) AND &H08) = 8 THEN GOTO WCTR2
          CALL READPIT(5,2,DLOW,DHIGH)
          TI(I + 1) = &H7FFF - (256 * DHIGH + DLOW) + 1
        NEXT I

        CALL SSTOP
        CLS
        PRINT "Time Results with TCK = "; TCK!; " Hz"
        PRINT "Time Interval in Sec,Elapsed Time in Sec"
        ETIME! = 0
        FOR I = 1 TO NEVT
          INTERVAL! = TI(I) * PTCK!
          ETIME! = ETIME! + INTERVAL!
          PRINT INTERVAL!; " , "; ETIME!
        NEXT I

ABORT: INPUT "Enter R to Repeat, Else Program Ends "; A$
        IF A$ = "R" OR A$ = "r" THEN GOTO MAIN

        END
```

The subroutines SSTOP, START, PROGPIT, LOADPIT and READPIT follow but are not reproduced here.

Estimation of the Shortest Time Between Events

The test program can be used to estimate the shortest time between events which can be successfully determined. Connect a variable frequency digital clock set initially on 10 Hz (period .100 second) to the EVENTS input. Use a frequency meter or oscilloscope to verify the period. Run the program. Select range 1 and NEVT = 2 or 4. The measured times should correspond to the known period. Increase the signal's frequency and repeat the measurement until the values are wrong. This occurs when the next timing interval starts while data is still being read.

9.3.4 Timer Capabilities and Limitations

The timer's capabilities and limitations are:
(1) The shortest time which can be measured depends on what operations are carried out in the data collection loop and on the speed of the computer and language. A typical value for PC-AT's using QuickBASIC is 1 ms.
(2) The longest time between two events is greater than 100,000 seconds (when TCK has the lowest possible frequency).
(3) Accuracy for a given timing interval is ±1 count which corresponds to ± one TCK cycle.
(4) The maximum number of timing intervals and therefore events is limited by the memory available for data storage.
(5) Finally, the events signal goes to a 74LS74 flip-flop's clock input. For an event to produce a toggle, the signal must be low at least 25 ns before going high and must then remain high at least 25 ns.
    In addition to these capabilities and limitations, the following cautions must be observed.
(1) There may be no warning if more than 32K TCK cycles occur in a timing interval.
(2) If the time between events is so short that there are no TCK cycles, then erroneous results occur.

9.4 Voltage Measurer

The CMS module contains circuitry which determines the voltage of one or two signals with 8- or 10-bit accuracy either at a programmable rate or upon an external trigger. Example measurements are: (1) a single voltage repeatedly determined and displayed (like a voltmeter); (2) two temperatures once every 10 ms for 100 seconds; (3) the light intensity of a single slit diffraction pattern as a function of the displacement of the detector across the pattern;

and (4) the energy spectrum of detected radiation (using the external trigger capability). In these and all other cases, information generated by sensors must be amplified to be between 0 and 10 volts.

The AD573 analog-to-digital converter was presented in Chapter 4, read device software in Chapter 7 and the ADC module in Chapter 8. Information from all these sources is needed to design and operate the CMS voltage measurer. Section 9.4.1 shows how to determine voltage at a programmable rate. Section 9.4.2 adds initiation and synchronization circuitry. Section 9.4.3 presents the external trigger. Section 9.4.4 covers test software. Section 9.4.5 summarizes capabilities and limitations. And Section 9.4.6 gives an example application.

9.4.1 Programmable Rate Voltage Measurement

PIT-C and the ADC module (with AD573-A and B) are the heart of programmable rate voltage measurement. The plan is for CTR_{C0} and CTR_{C1} to produce a Voltage ClocK (VCK) which initiates a conversion once each cycle. Specifically, when VCK goes high, CTR_{C2} generates a Convert Pulse (CV) which causes both ADC's to determine voltage. As shown in Figure 4.5, ADC-DR* (ADC-A's Data Ready) goes high just after the falling edge of CV and returns low when data is ready. The latter event clears a flip-flop whose

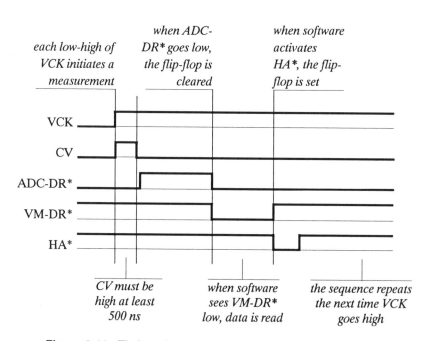

Figure 9.11. Timing diagram for a voltage measurement. The sequence is initiated by VCK going high and ends with HA*'s activation by software.

output is VM-DR* (Voltage Measurer Data Ready). Software monitors VM-DR* and reads data when clear is found. In doing so, **HA*** (High byte enable of ADC-A) is activated which sets the flip-flop for the next conversion. The scheme is shown in Figure 9.11.

Figure 9.12 gives circuitry which implements the plan. The 1.0000 MHz digital oscillator goes to CK_0. G_0 is wired high. OUT_0 is the Base ClocK (BCK) introduced in Section 9.1.1. It goes to CK_1 as well as to the counter and timer circuits. G_1 is involved in initiation as discussed in the next section. OUT_1 is VCK which goes to G_2. Both CTR_{C0} and CTR_{C1} are programmed for mode 3. The loaded values determine the measurement rate in a way identical to the counting interval frequency discussed in Section 9.2.1. An example is given in Section 9.4.4.

CTR_{C2} is programmed for mode 1 and generates CV. CK_2 is the 1.0000 MHz clock. When G_2 goes high (once each VCK cycle), OUT_2 goes low for a length of time equal to CTR_2's loaded value multiplied by the period of the signal at CK_2 (as originally presented in Section 6.2.1). So, if the value is 1,

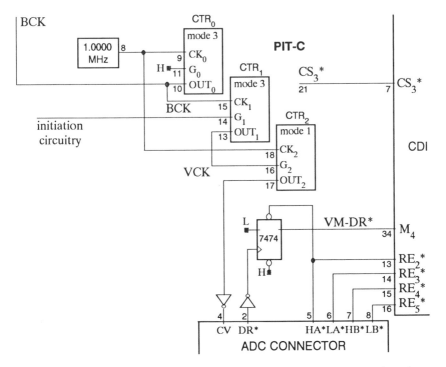

Figure 9.12. Circuitry which carries out programmable rate voltage measurement. Relevant CDI and ADC connections are shown. Not shown are PIT control lines (WR*, RD*, A_0 and A_1) and CDI data lines to the PIT and to the ADC module.

OUT_2 goes low 1 μsec. CV in Figure 9.11 is the inverse of OUT_2 and goes to the ADC connector as shown in Figure 9.12.

ADC-DR* goes low when data is ready. Therefore, its inverse goes high and clears the flip-flop (via the clock/data mode). When the flip-flop output's VM-DR* (connected to the CDI's M_4) goes low, software responds by activating RE_2* (via CALL READD(2,HIGHD)). With RE_2* connected to the flip-flop's SET* and to the ADC module's HA*, the action causes 8 bits to be put on the data lines and returns VM-DR* high. Finally, RE_3* goes to the ADC module's LA*, RE_4* to HB* and RE_5* to LB*. These controls are used by software depending on whether one or two voltages are measured and as necessary for the desired accuracy (8 or 10 bits).

The following PIT to CDI connections are not shown in the drawing: WR* to CO_1, RD* to CO_2, A_0 to CO_6, A_1 to CO_7, and D7...D0 to $d_7...d_0$. Also, CDI's $d_7...d_0$ is connected to the ADC's $d_7...d_0$.

9.4.2 Voltage Measurement Initiation

Circuitry in Figure 9.13 enables software and hardware to initiate voltage measurement. The two AND gates and two flip-flops play the same roles as in the counter circuit in Section 9.2.2. V-S/S* (Voltage Start/Stop) goes high when both S-S/S* and EX-S/S* become high. However, SYNC keeps VCK low until the beginning of the first rate period (which occurs at the end of a

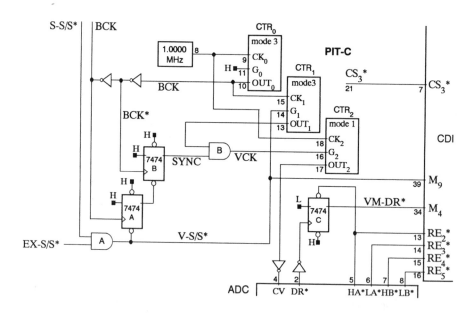

Figure 9.13. Additional circuitry which carries out initiation and synchronization.

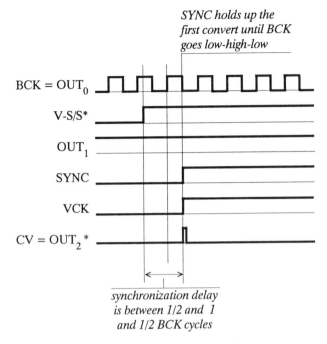

SYNC holds up the first convert until BCK goes low-high-low

synchronization delay is between 1/2 and 1 and 1/2 BCK cycles

Figure 9.14. Timing diagram showing initiation and synchronization. V-S/S* goes high when both S-S/S* and EX-S/S* become high. Flip-flops A and B delay V-S/S* going high until BCK goes low-high-low.

BCK low-high-low after V-S/S* goes high). Initiation and synchronization are summarized in Figure 9.14.

V-S/S* also goes to the CDI's M_9 allowing software to monitor for external abort (if M_7 and M_8 are low).

BCK goes through an inverter and BCK* goes to FF-B. BCK* is inverted back to BCK which goes on to FF-A and to the counter and timer circuits. The second inverter is logically unnecessary but improves fanout (since BCK goes to numerous gates in the other circuits).

9.4.3 External Trigger

It's desirable to be able to initiate voltage measurement when the experimental hardware supplies a trigger. The issue is how to add the operation while maintaining programmable rate capability.

Figure 9.15 gives the circuit. EX-TR (**EX**ternal **TR**igger) must be low when not used. This makes AND-C and therefore AND-D low. With one input low, the OR output equals the other input which is VCK. So, programmable rate measurement occurs as described above.

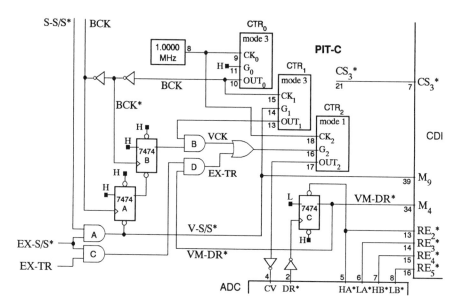

Figure 9.15. The complete voltage measurer circuit including external trigger. When EX-TR is held low, VCK goes to G_2. When S-S/S* stays low, EX-TR goes to G_2.

When the trigger is used, software stop is maintained which keeps S-S/S* and V-S/S* low. This makes AND-A, both flip-flops and AND-B all low. With AND-B low, the OR output is EX-TR which goes to G_2. So a conversion occurs every time EX-TR goes from low to high. (VM-DR* and software operate the same as before.)

EX-S/S* works in the trigger mode. When low, AND-C is low and EX-TR does not reach G_2.

Since external triggers may occur randomly (as in the case of radioactive decay), EX-TR might go high before data from the previous conversion is read. To prevent this, EX-TR and VM-DR* are connected to AND-D and the result goes on to G_2. Since VM-DR* is low when data is ready but not read, AND-D is low and EX-TR does not reach G_2.

In the case described here, triggers which occur while VM-DR* is low are lost. A more sophisticated approach is to use a separate flip-flop which is cleared when CV goes high and set at the end of data read. Also, elaborate schemes which minimize lost triggers are possible. However, there is a limit on how fast converted values can be processed and therefore on the maximum rate of acquisition (as disucssed in the next section).

9.4.4 Voltage Software

Programmable rate voltage measurement is initiated and controlled by the test/demonstration program developed in this section. The goal is to repeatedly determine two voltages at 10-bit accuracy with both the rate and number of repetitions selectable. The starting point is the modified software shell CMS.BAS introduced in Section 9.2.3.

Constants and Arrays

The voltage measurer requires the three counters in PIT-C. The following table gives each counter's mode, index, control word, LOW and HIGH load bytes and function.

counter	mode	index	CW	LOW	HIGH	function
C0	3	3,0	&H36	X	X	generate BCK
C1	3	3,1	&H76	X	X	generate VCK
C2	1	3,2	&HB2	1	0	generate CV

"X" means the value is determined later. As previously discussed, CTR_{C2} is programmed for mode 1 and loaded with 1.

Arrays are needed for LOW, HIGH and CW. Also, arrays are needed to store data from Voltage CHannel's **A** and **B**. The declarations are:

```
DIM LOW(7,3),HIGH(7,3),CW(7,3)
DIM VCHA(1000),VCHB(1000)
```

One thousand is picked for the maximum number of measurements (which can be increased or decreased as needed).

These constants are assigned values from the table and go in CMS.BAS.

```
CW(3,0) = &H36
CS(3,1) = &H76
CW(3,1) = &HB2 : LOW(3,2) = 1 : HIGH(3,2) = 0
```

Since PIT's A and B (CS_5^* and CS_6^*) are not involved in voltage measurement, their constants are deleted.

Initialization

All CMS programs begin with:

```
MAIN: CALL SSTOP
      CLS
```

The next task is to set the measurement rate. The plan here is for the rate to be selectable between 0.10 and 1000 measurements/second. To achieve this range, BCK's frequency is set and left at 5000 Hz. CTR_{C0}'s load value (1,000,000/BCK) is 200 which makes:

```
LOW(3,0) = 200 : HIGH(3,0) = 0
```

With BCK equal to 5000 Hz, VCK's frequency is 5000/[value loaded into CTR_{C1}]. For 1000 Hz, value = 5; for 10 Hz, value = 500; and for 0.10 Hz, value = 50,000. The following program statements input a measurement rate and then compute value, LOW and HIGH.

```
PRINT "Enter Meas/Sec: Max = 1000, Min = .1"
INPUT RATE!
VALUE! = 5000/RATE!
HIGH(3,1) = INT(VALUE!/256)
LOW(3,1) = INT(VALUE!) - (256 * HIGH(3,1))
RATE! = 5000/(HIGH(3,1) * 256 + LOW(3,1))
RPERIOD! = (HIGH(3,1) * 256 + LOW(3,1))/5000
PRINT "Actual Rate: "; RATE!; " Meas/Sec"
```

The real variables are: RATE! (**RATE** of voltage conversion in measurements/second); VALUE! (**VALUE** used in computing high and low load bytes); RPERIOD! (**Rate PERIOD** = 1/RATE is the time between measurements).

The usual method is used to calculate HIGH and LOW. For instance, if an "even" 10 Hz is entered for RATE!, VALUE! = 500, HIGH(3,1) = 1 and LOW(3,1) = 244. On the other hand, if 800 Hz is entered, VALUE! = 6.25, HIGH(3,1) = 0 and LOW(3,1) = 6. In this case, six is the (necessarily integer) load value. This makes the actual rate 5000/6 = 833.3 Hz. Because the selected and programmed rates may be different, after LOW and HIGH are determined, the actual rate is computed and displayed. RPERIOD! is used later to determine elapsed times.

In general, there is a difference between the selected and actual measurement rates anytime the former is not an even divisor of 5000 (BCK). In such cases, the larger the selected rate, the larger the potential difference. If greater accuracy is required, BCK can be increased. However, this leads to a smaller range of rates. Of course, more elaborate software allows BCK to be optimized for whatever selection is made.

The other input parameter is NVR (the total Number of Voltage Readings).

```
INPUT "Enter the Number of Voltage Readings ";NVR
```

PIT-C's counters are programmed and loaded as follows:

```
FOR J = 0 TO 2
  CALL PROGPIT(3,CW(3,J))
  CALL LOADPIT(3,J,LOW(3,J),HIGH(3,J))
NEXT J
```

Data Collection Loop

After sending start, software waits for M_4 to go low or, preferably, waits while M_4 is high. FF-C in Figure 9.15 originates M_4. It is cleared by data ready and set by RE_2^*. A problem is the state of the flip-flop before the first read. To be sure it's set, read device 2 is called before data acquisition starts.

```
CALL READD(2,HIGHD)
CALL START
```

HIGHD is the **HIGH** 8 **D**ata bits of a converted value which is meaningless in this case.

The following statements implement the data collection loop.

```
        FOR I = 1 TO NVR
WFLOW:   IF (INP(CP) AND &H10) = 16 THEN GOTO WFLOW
         CALL READD(2,HIGHD)
         CALL READD(3,LOWD)
         VCHA(I) = 4 * HIGHD + (LOWD AND &H03)
         CALL READD(4,HIGHD)
         CALL READD(5,LOWD)
         VCHB(I) = 4 * HIGHD + (LOWD AND &H03)
         IF INKEY$ <> "" THEN GOTO REPT
        NEXT I
DSPLY: CALL SSTOP
```

The WFLOW (Wait For **LOW**) statement repeats as long as M_4 is high. Specifically, INP(CP) obtains from the control-in port X X X M4 X X X X which after ANDing with 00010000 (16) equals 0 0 0 M4 0 0 0 0. If M_4 is high, the result is 10 hex or 16; and if M_4 is 0, the result is 0.

When wait ends, both the high 8 and low 2 bits from each ADC are read. (LOWD stands for the **LOW** 2 **D**ata bits). Recall from Section 4.6 that the low data bits must be masked with 0000011 or 03 hex and the 10-bit converted value is 4*HIGHD + LOWD.

Data collection is aborted by typing any key which causes execution to branch to the repeat option (given below).

The case of reading 10 bits from both ADC's makes a "longer" data collection loop and therefore reduces the maximum rate of acquisition. Clearly, reading only the high 8 bits from one channel is faster. The loop above executes in less than 1 ms when the program is in a QuickBASIC .EXE standalone form running on a 12.5 MHz PC-AT compatible. A general method to determine the maximum rate is given later.

Display Results and Repeat

The following program block displays the voltage from both channels and the elapsed time.

```
DSPLY: CALL SSTOP
       CLS
       PRINT "Results of Voltage Measurements"
       PRINT "Format: CH-A Value,CH-B Value,Net Time "
       CONV! = 10/1024
```

```
FOR I = 1 TO NVR
  PRINT USING "##.##"; VCHA(I) * CONV!;
  PRINT " , ";
  PRINT USING "##.##"; VCHB(I) * CONV!;
  PRINT " , ";
  PRINT (I-1) * RPERIOD!
NEXT I
```

Recall from Section 4.1 that voltages are computed as follows:

```
VOLTS! = (10-bit value/1024) * 10
```

CONV! (CONVersion constant) equals 10/1024. Voltages are displayed to two decimal places since 10-bit accuracy is to the nearest .01 volt (as discussed in Section 4.1). If measurement 1 occurs at $t = 0$ and if measurements are RPERIOD seconds apart, the elapsed time of the Ith measurement is (I-1)*RPERIOD!.

The repeat option is the same as in other CMS programs.

Voltage Test/Demonstration Program

The bits and pieces above when integrated into CMS.BAS produce the following complete program.

```
'   Program:  VOLTAGE
DEFINT A-Z
DIM SELPP(7),PROGP(7)
DIM SELPC(7,3),LOADPC(7,3),READPC(7,3)
DIM REDD(6)
DIM LOW(7,3),HIGH(7,3),CW(7,3)
DIM VCHA(1000),VCHB(1000)
COMMON SHARED CP,DP,NORM,STRT,SSTP
COMMON SHARED SELPP(),PROGP(),SELPC()
COMMON SHARED LOADPC(),READPC()
COMMON SHARED REDD()
CP = &H300 : DP = &H301
NORM = &HFF
STRT = &HC6 : SSTP = &HCE
CW(3,0) = &H36 : CW(3,1) = &H76 : CW(3,2) = &HB2
LOW(3,0) = 200 : LOW(3,2) = 1
HIGH(3,0) = 0 : HIGH(3,2) = 0
SELPP(3) = &HDF
PROGP(3) = &HDD
SELPC(3,0) = &H1F : SELPC(3,1) = &H5F
SELPC(3,2) = &H9F
LOADPC(3,0) = &H1D : LOADPC(3,1) = &H5D
LOADPC(3,2) = &H9D
READPC(3,0) = &H1B : READPC(3,1) = &H5B
READPC(3,2) = &H9B
```

```
          REDD(2) = &HD6 : REDD(3) = &HDE
          REDD(4) = &HE6 : REDD(5) = &HEE
MAIN:  CALL SSTOP
       CLS
       PRINT "Enter Meas/Second: Max = 1000, Min = .1"
       INPUT RATE!
       VALUE! = 5000/RATE!
       HIGH(3,1) = INT(VALUE!/256)
       LOW(3,1) = INT(VALUE!) - (256 * HIGH(3,1))
       RATE! = 5000/(HIGH(3,1) * 256 + LOW(3,1))
       RPERIOD! = (HIGH(3,1) * 256 + LOW(3,1))/5000
       PRINT "Actual Rate: "; RATE!; " Meas/Second"
       INPUT "Enter the Number of Voltage Meas "; NVR
       FOR J = 0 TO 2
          CALL PROGPIT(3,CW(3,J))
          CALL LOADPIT(3,J,LOW(3,J),HIGH(3,J))
       NEXT J
       CALL READD(2,HIGHD)
       CALL START
       FOR I = 1 TO NVR
WFLOW:    IF (INP(CP) AND &H10) = &H10 THEN GOTO WFLOW
          CALL READD(2,HIGHD)
          CALL READD(3,LOWD)
          VCHA(I) = 4 * HIGHD + (LOWD AND &H03)
          CALL READD(4,HIGHD)
          CALL READD(5,LOWD)
          VCHB(I) = 4 * HIGHD + (LOWD AND &H03)
          IF INKEY$ <> "" THEN GOTO REPT
       NEXT I
DSPLY: CALL SSTOP
       CLS
       PRINT "Results of Voltage Measurements"
       PRINT "Format  CH-A Value, CH-B Value, Net Time"
       CONV! = 10/1024
       FOR I = 1 TO NVR
          PRINT USING "##.##"; VCHA(I) * CONV!;
          PRINT " , '';
          PRINT USING "##.##"; VCHB(I) * CONV!;
          PRINT " , ";
          PRINT (I-1) * RPERIOD!
       NEXT I
REPT:  INPUT "Enter R to Repeat, Else Program Ends "; A$
       IF A$ = "R" OR A$ = "r" THEN GOTO MAIN
       END
```

The subroutines SSTOP, START, PROGPIT, LOADPIT, READPIT and READD follow but are not reproduced here.

Just as for the counter and timer test/demonstration programs, the intention here is to demonstrate the basic interactions between software and hardware. Limitless variations are possible. Also, real time graphics, elaborate menus, file I/O and data computations can be introduced as desired.

Estimation of the Maximum Measurement Rate

The test program (and its variations) can be used to roughly determine the maximum measurement rate as follows. Let one of the voltage inputs be a known frequency digital clock. The measured values should jump from 0 (or so) to 5 (or so) once each clock cycle. Run the program and select 100 Hz for the rate and 25 for the number of measurements. If the clock's frequency is 1000 Hz, there should be 10 measurements between 0 to 5 jumps. Increase the rate to 500 Hz and the clock to 5000 Hz. Again, there should be 10 measurements between jumps. Repeat until the expected number of measurements is **NOT** found which occurs when software can't keep up with conversions.

9.4.5 Voltage Measurer Capabilities and Limitations

The capabilities and limitations of programmable rate voltage measurement are:

(1)   The fastest rate depends on what operations are carried out in the data collection loop and on the speed of the computer and language. A typical value for PC-AT's using QuickBASIC is 1000 measurements/second.

In the external trigger mode (illustrated in the next section), the minimum time between triggers corresponds to the maximum rate.

It's unlikely that greater rates could be achieved by faster ADC's. The AD573's 20 μsec conversion time is small compared to the time software requires to read values. A method to achieve much faster rates is described in Chapter 10.

(2)   The slowest rate is over 2500 seconds between conversions.

(3)   Measured voltages must be between 0 and 10.

(4)   Accuracy for 10-bit conversion is $\pm.01$ volts; accuracy for 8 bits is $\pm.04$ volts (as discussed in Section 4.1).

(5)   The maximum number of measurements depends on available memory.

(6)   According to specifications for 8253 gate inputs, the external trigger must be low at least 100 ns before going high and must stay high at least 150 ns.

In addition to these capabilities and limitations, the following cautions must be observed:

(1) There is no warning (in terms of incorrect voltage values) if the measurement rate is too fast. The possibility is dealt with by knowing the maximum rate for the computer and software being used.

(2) Errors may occur if the voltage being measured "changes" during the 20 μsec AD573 conversion time. Rapidly changing voltages require a sample and hold such as Analog Devices' model AD582.

9.4.6 Example Application

The test/demonstration program in Section 9.4.4 determines voltages from two signals at a selectable rate. The objective of this section is to demonstrate the external trigger operation and give an example of voltage measurement with a different interpretation of results.

A voltage analyzer counts the number of times each voltage in a range is measured. The plan here is to repeatedly read the high 8 bits converted by ADC-A and add one to an array element which corresponds to the value. Specifically, the 256 possible 8-bit values (00000000 to 11111111) divide the ADC's 0 to 10 volt range into 256 increments each slightly less than .04 volt. Array elements correspond to measured voltages as follows:

voltage range	converted value	array index
.00 to .04	00000000	0
.04 to .08	00000001	1
.08 to .12	00000010	2
.12 to .16	00000011	3
...	...	...
9.92 to 9.98	11111110	254
9.96 to 10.00	11111111	255

(The ranges above are approximate since $10/256 = .0391$, not .04.)

When data is ready, the following program steps add 1 to the appropriate element. VANA (Voltage ANAlyzer) is the array.

```
CALL READD(2, HIGHD)
VANA(HIGHD) = VANA(HIGHD) + 1
```

After acquiring data, VANA(I) equals the number of times I was the converted value and therefore the number of times the voltage corresponding to I was measured.

An external trigger initiates the conversions which are read and analyzed. Only PIT-C's CTR_2 is needed for the operation. Specifically, CTR_{C2} generates a convert pulse when its gate (connected to EX-TR) goes high. The following table gives the counter's mode, index, control word and load bytes.

counter	mode	index	CW	LOW	HIGH	function
C2	1	3,2	&HB2	1	0	generate CV

The software starting point is CMS.BAS. The additional declarations and constants are:

```
DIM LOW(7,3),HIGH(7,3),CW(7,3)
DIM VANA(256)
CW(3,2) = &HB2 : LOW(3,2) = 1 : HIGH(3,2) = 0
```

The complete program is given below. The first operation is CALL SSTOP. Next, the PIT counter is programmed and loaded. The data array VANA is zeroed. After "type any key to start," READD(I,HIGHD) is called to initialize the flip-flop as discussed earlier. The wait for low loop is identical to the one in the test/demonstration program. The IF INKEY statement provides an exit to an otherwise infinite data collection loop. Afterward, values are displayed 12 to a row. With unnecessary constants and subroutines removed, the program is:

```
'       Program:  VOLTAGE ANALYZER
        DEFINT A-Z
        DIM SELPP(7),PROGP(7)
        DIM SELPC(7,3),LOADPC(7,3),READPC(7,3)
        DIM REDD(6)
        DIM LOW(7,3), HIGH(7,3), CW(7,3)
        DIM VANA(256)
        COMMON SHARED CP,DP,NORM,STRT,SSTP
        COMMON SHARED SELPP(),PROGP(),SELPC()
        COMMON SHARED LOADPC(),READPC()
        COMMON SHARED REDD()
        CP = &H300: DP = &H301
        NORM = &HFF
        STRT = &HC6: SSTP = &HCE
        CW(3,2) = &HB2 : LOW(3,2) = 1 : HIGH(3,2) = 0
        SELPP(3) = &HDF
        PROGP(3) = &HDD
        SELPC(3,2) = &H9F
        LOADPC(3,2) = &H9D
        READPC(3,2) = &H9B
        REDD(2) = &HD6
MAIN:   CALL SSTOP
        CALL PROGPIT(3,CW(3,2))
        CALL LOADPIT(3,2, LOW(3,2), HIGH(3,2))
        FOR I = 1 TO 256
          VANA(I) = 0
        NEXT I
        CLS
        PRINT "Voltage Analyzer"
        INPUT "Any Key to Start, Then Any to Stop "; A$
        CALL READD(2,HIGHD)
```

```
WFLOW: IF (INP(CP) AND &H10) = &H10 THEN GOTO WFLOW
       CALL READD(2,HIGHD)
       VANA(HIGHD) = VANA(HIGHD) + 1
       IF INKEY$ <> "" THEN GOTO DSPLY
       GOTO WFLOW

DSPLY: CLS
       PRINT "Voltage Analysis Results"
       FOR I = 1 TO 252 STEP 12
         FOR J = 0 TO 11
            PRINT USING "#####"; VANA(I + J);
         NEXT J
         PRINT
       NEXT I

       PRINT
REPT:  INPUT "Enter R to Repeat, Else Program Ends "; A$
       IF A$ = "R" OR A$ = "r" THEN GOTO MAIN

       END
```

Subroutines SSTOP, START, PROGPIT, LOADPIT, READPIT and READD follow but are not reproduced here.

Recall from the voltage measurement circuitry that EX-TR passes to G_2 only if software stop is maintained. Therefore, start is never called.

There are two distinct ways to operate the analyzer. In one, EX-TR is a digital clock and voltage conversion is periodic. Except for data interpretation and the clock's origin, this is no different from programmable rate measurement.

In the second approach, events (such as the detection of radioactive decay) control when conversions occur. For example, a pulse height analyzer is created by supplying ADC-A with pulses whose voltages (heights) are proportional to the energy of detected radiation and by supplying EX-TR with a digital signal which goes high at the start of each pulse. Array elements in the resulting data contain the number of times radiation with the corresponding energy was detected. Support electronics must keep each pulse constant for the 20 μsec needed for conversion. A sample and hold (such as the Analog Devices AD582) is usually required.

There are several ways to enhance the analyzer's capabilities. With a single precision integer array, the maximum number of counts is 32K. Capacity is significantly increased by using double precision integer or real arrays. By reading all 10 ADC bits, voltage is analyzed in 1024 increments.

The maximum rate of analysis depends on what operations are in the acquisition loop and the speed of the computer and language. The following procedure allows the rate to be estimated. Connect a variable frequency digital clock set initially on 100 Hz to EX-TR. Apply a DC voltage to ADC-A. Run the program and acquire data for 10 seconds (on a stopwatch). One

thousand counts should occur in the one or two array elements corresponding to the voltage. Increase the frequency to 500 Hz and see if 5000 counts
occur in 10 seconds. Repeat the procedure until expectation is not met which
occurs when triggers are missed.

9.5 Complete CMS Circuit

Figure 9.16 is the complete CMS circuit. The top third is the counter from
Figure 9.8, the middle third is the timer from Figure 9.10 and the bottom
third is the voltage measurer from Figure 9.15. The CDI connector is on the
right and the ADC is on the bottom. The six input signals are on the left.
(Voltages A and B go to the ADC module and are not shown.) The PIT
controls (WR*, RD*, A_0 and A_1) and CDI data lines are not shown. The
AND's, INVERTER's and flip-flops in the diagram are labeled A1, C2, etc.
The designations are used in Sections 9.7 and 9.8.

9.6 Example of Integrated Measurement

So far, counting, timing and voltage measurements have been individually
described. The purpose of this section is to illustrate measurement combinations by solving a data acquisition problem which requires a sequence of time
and voltage determinations.

9.6.1 Data Acquisition Problem

Consider the situation in Figure 9.17. An experiment produces a sequence of
four pulses. The objective is to measure the time between the first two, voltage and time from the second to the third and then time again between the
third and fourth.

> *This data acquisition scenario might arise from the following mechanics
> experiment. The first two pulses occur as an air track glider passes a pho-
> togate heading for a collision with a "bumper." The second two pulses
> occur after the collision as the glider passes the photogate going the other
> way. The voltage measured between the second and third pulses origi-
> nates from a strain gauge attached to the bumper.*

9.6.2 Solution

Suppose the pulses go to the timer's EVENTS input. The gates of PIT-B's
CTR_1 and CTR_2 then follow the pattern in Figure 9.18. G_1 goes to the CDI's
M_2 (which is control-in bit 2) and G_2 goes to M_3 (CI_3). Suppose also that the
voltage to be measured goes to ADC-A and PIT-C is set up for programmable rate measurement. Recall that VM-DR* (M_4) goes low at the end of

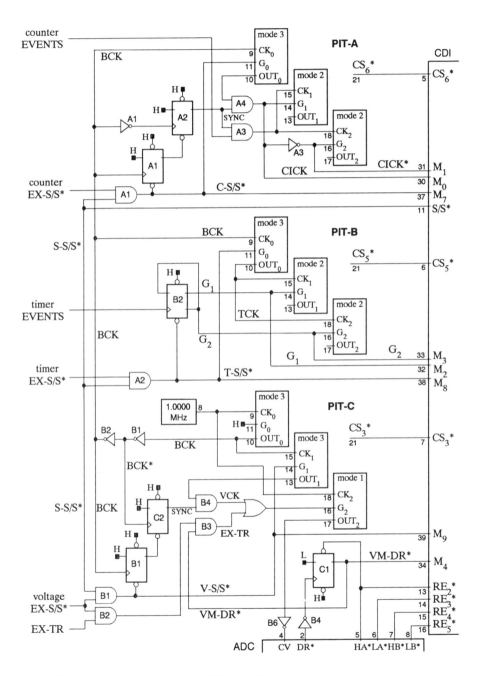

Figure 9.16. The complete CMS circuit.

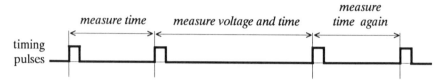

Figure 9.17. Timing diagram for a sequence of time and voltage measurements.

conversions and returns high when software reads the high byte from ADC-A.

The following sequence makes the desired measurements:
(1) Software makes S-S/S* high in both the timer and voltage circuits. (The EX-S/S*'s are not used.)
(2) G_1 is low until the first timing pulse. Therefore, software waits while G_1 is low.
(3) The first pulse starts CTR_1 and software waits while G_1 is high.
(4) The second pulse stops CTR_1 and starts CTR_2. The previous wait ends and data is read from CTR_1.
(5) Next, software enters a loop which continues until the third pulse takes G_2 low. Each time around the loop, software checks G_2, waits while VM-DR* is high and then reads the high byte from ADC-A.
(6) The third pulse stops CTR_2 and starts CTR_1 again. The voltage loop ends and data is read from CTR_2.
(7) Software waits while G_1 is high.
(8) The fourth pulse takes G_1 low and stops CTR_1. The previous wait ends and software reads the final data from CTR_1.

These program steps carry out the sequence.

```
NVR = 0
CALL START
```

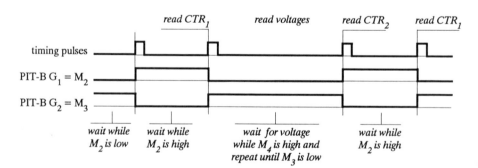

Figure 9.18. Data acquisition plan for time and voltage measurements.

```
WFP1:  IF (INP(CP) AND &H04) = 0 THEN GOTO WFP1
WFP2:  IF (INP(CP) AND &H04) = 4 THEN GOTO WFP2
       CALL READPIT(5,1,LD1,HD1)
WFP3:  IF (INP(CP) AND &H08) = 0 THEN GOTO GCTR2
WFV:   IF (INP(CP) AND &H10) = 16 THEN GOTO WFV
       CALL READD(2,CHA(NVR))
       NVR = NVR + 1
       GOTO WFP3
GCTR2: CALL READPIT(5,2,LD2,HD2)
WFP4:  IF (INP(CP) AND &H04) = 4 THEN GOTO WFP4
       CALL READPIT(5,1,LD3,HD3)
       CALL SSTOP
```

WFP1 (Wait For Pulse 1) loops while $G_1 = M_2$ is low. The mask is 00000100 or 4. WPF2 (Wait For Pulse 2) loops while M_2 is high. LD1 and HD1 are the Low and High Data from CTR_1.

WFP3 (Wait For Pulse 3) starts the voltage measurement loop. If $G_2 = M_3$ is low, the loop ends and the program branches to GCTR2 (Get data from CTR_2). If M_3 is high, WFV (Wait For Voltage) loops while VM-DR* $= M_4$ is high. When voltage is ready, device 2 (high byte of ADC-A) is read and the value stored in the array CHA (CHannel A). NVR (Number of Voltage Readings) counts voltage measurements and supplies the array index. After each reading, execution branches back to WFP3. The mask for M_3 is 00001000 (8) and for M_4 is 00010000 (16).

GCTR2 reads and stores CTR_2 data in LD2 and HD2. Finally, WFP4 (Wait For Pulse 4) loops while M_2 is high. Then CTR_1 is read and values given to LD3 and HD3.

As soon as software executes start, the voltage clock (VCK) produces conversions. However, it's not until pulse 2 that software looks at VM-DR* and inputs the most recent converted value.

The remaining issues are what timing clock (TCK) frequency and voltage rate to use, and how to compute voltage and time values.

9.6.3 Constants and Computations

VCK controls the rate of conversion and TCK determines the range and accuracy of time measurements. Both are produced by dividing the base clock (BCK) by the values loaded in CTR_{C1} and CTR_{B0}, respectively. For the case here, TCK is picked at 10 KHz. This means times as short as .01 second can be measured with 1% accuracy. The voltage rate is picked at 500 measurements/second. This yields 50 readings in .10 second. To support both TCK and VCK, 100 KHz is picked for BCK. To operate the timer and voltage measurer, PIT B (CS_5^*) and PIT C (CS_3^*) must be programmed and loaded. The following table gives all six counters' mode, index, mode control word, low and high load bytes and function.

counter	mode	index	CW	LOW	HIGH	function
C0	3	3,0	&H36	10	0	generate BCK
C1	3	3,1	&H76	200	0	generate VCK
C2	1	3,2	&HB2	1	0	generate CV
B0	3	6,0	&H36	10	0	generate TCK
B1	2	6,1	&H74	&HFF	&H7F	count TCK cycles
B2	2	6,2	&HB4	&HFF	&H7F	count TCK cycles

The following computations are made on values obtained during data acquisition:

(1) The Number of **TCK** cycles in the first interval is:

```
NTCK1 = &H7FFF - (256 * HD1 + LD1) + 1
```

Time **INT**erval **1** in seconds equals the number of TCK cycles multiplied by TCK's period (1/10000). So:

```
TINT1! = NTCK1/10000
```

(2) Similarly, Time **INT**erval **2** (TINT2) between pulses 2 and 3 is computed from HD2 and LD2.

(3) And TINT3 between pulses 3 and 4 is computed from HD3 and LD3.

(4) The net Time of Voltage Measurement is NVR multiplied by VCK's period:

```
TVM! = NVR * (1/500).
```

TVM should equal TINT2

(5) The array CHA has NVR values. Each is the high 8 bits of a converted value. Voltage is computed as discussed in Section 4.1:

```
CONV! = 10/256
VOLTS(I) = CHA(I) * CONV!
```

9.6.4 The Program

Starting with the shell CMS.BAS in Section 9.2.3, the following program carries out the sequence of time and voltage measurements.

```
'        Program:  TIME-VOLTAGE
DEFINT A-Z
DIM SELPP(7),PROGP(7)
DIM SELPC(7,3),LOADPC(7,3),READPC(7,3)
DIM REDD(6)
DIM LOW(7,3),HIGH(7,3),CW(7,3)
DIM CHA(1000)
COMMON SHARED CP,DP,NORM,STRT,SSTP
COMMON SHARED SELPP(),PROGP(),SELPC()
COMMON SHARED LOADPC(),READPC()
COMMON SHARED REDD()
```

```
          CP = &H300 : DP = &H301
          NORM = &HFF
          STRT = &HC6 : SSTP = &HCE
          CW(5,0) = &H36 : LOW(5,0) = 10 : HIGH(5,0) = 0
          CW(5,1) = &H74 : LOW(5,1)= &HFF : HIGH(5,1)= &H7F
          CW(5,2) = &HB4 : LOW(5,2)= &HFF : HIGH(5,2)= &H7F
          SELPP(5) = &HEF
          PROGP(5) = &HED
          SELPC(5,0) = &H2F : SELPC(5,1) = &H6F
          SELPC(5,2) = &HAF
          LOADPC(5,0) = &H2D : LOADPC(5,1) = &H6D
          LOADPC(5,2) = &HAD
          READPC(5,0) = &H2B : READPC(5,1) = &H6B
          READPC(5,2) = &HAB
          CW(3,0) = &H36 : LOW(3,0) = 10 : HIGH(3,0) = 0
          CW(3,1) = &H76 : LOW(3,1) = 200 : HIGH(3,1) = 0
          CW(3,2) = &HB2 : LOW(3,2) = 1 : HIGH(3,2) = 0
          SELPP(3) = &HDF
          PROGP(3) = &HDD
          SELPC(3,0) = &H1F : SELPC(3,1) = &H5F
          SELPC(3,2) = &H9F
          LOADPC(3,0) = &H1D : LOADPC(3,1) = &H5D
          LOADPC(3,2) = &H9D
          READPC(3,0) = &H1B : READPC(3,1) = &H5B
          READPC(3,2) = &H9B
          REDD(2) = &HD6
MAIN:     CALL SSTOP
          CLS
          FOR J = 0 TO 2
            CALL PROGPIT(3,CW(3,J))
            CALL PROGPIT(5,CW(5,J))
            CALL LOADPIT(3,J,LOW(3,J),HIGH(3,J))
            CALL LOADPIT(5,J,LOW(5,J),HIGH(5,J))
          NEXT J
          NVR = 0
          INPUT "Type Any Key to Start Acquisition "; A$
          CALL READD(2,HIGHD)
          CALL START
WFP1:     IF (INP(CP) AND &H4) = 0 THEN GOTO WFP1
WFP2:     IF (INP(CP) AND &H4) = 4 THEN GOTO WFP2
          CALL READPIT(5,1,LD1,HD1)
WFP3:     IF (INP(CP) AND &H8) = 0 THEN GOTO GCTR2
WFV:      IF (INP(CP) AND &H10) = 16 THEN GOTO WFV
          CALL READD(2,CHA(NVR))
          NVR = NVR + 1
          GOTO WFP3
GCTR2:    CALL READPIT(5,2,LD2,HD2)
WFP4:     IF (INP(CP) AND &H4) = 4 THEN GOTO WFP4
          CALL READPIT(5,1,LD3,HD3)
```

```
         CALL SSTOP
         NTCK1 = &H7FFF - (256 * HD1 + LD1) + 1
         NTCK2 = &H7FFF - (256 * HD2 + LD2) + 1
         NTCK3 = &H7FFF - (256 * HD3 + LD3) + 1
         TINT1! = NTCK1 / 10000
         TINT2! = NTCK2 / 10000
         TINT3! = NTCK3 / 10000
         TVM! = NVR / 500
         PRINT "Result of Time and V(t) Measurements"
         PRINT "Time 1: "; TINT1!; " sec"
         PRINT "Time 2: "; TINT2!; " sec"
         PRINT "Time 3: "; TINT3!; " sec"
         PRINT "Time for "; NVR; " Vo Mea: "; TVM!; " sec"
         INPUT "Type Any Key for Voltage Values"; A$
         CONV! = 10 / 255
         FOR I = 0 TO NVR STEP 16
           FOR J = 0 TO 15
             PRINT USING "#.##"; CHA(I + J) * CONV!;
             PRINT " ";
           NEXT J
           PRINT
         NEXT I

REPT:    INPUT "Enter R to Repeat, Else Program Ends "; A$
         IF A$ = "R" OR A$ = "r" THEN GOTO MAIN

         END
```

Subroutines SSTOP, START, PROGPIT, LOADPIT, READPIT and READD follow but are not reproduced here.

The reason for CALL READD(2,HIGHD) before CALL START is to make sure VM-DR* is high before the first conversion.

The time between pulses cannot be less than the minimum discussed earlier, and the voltage measurement rate cannot exceed the maximum discussed earlier.

The sequence described here illustrates one of many possibilities. But the CMS supports measurements from only one count EVENTS signal, one time EVENTS signal and two voltages. When greater capability is needed, the techniques illustrated here are applicable to the design of a new measurement module. Several exercises at the end of this chapter explore possibilities. Also, the next chapter presents an approach for faster acquisition.

9.7 Construction

The CMS circuit may be constructed on a wire wrap card identical to the one used for the CDI and ADC modules. This section lists components, suggests a layout and gives construction guidelines. Section 9.8 outlines testing and troubleshooting.

9.7.1 CMS Components

The CMS circuit in Figure 9.16 requires the following components.
(1)   Three 8253-5 PIT's and three 24-pin wire wrap sockets.
(2)   Three 74LS74 flip-flops.
(3)   Two 74LS04 INVERTER'S, two 74LS08 AND'S and one 74LS32 OR.
(4)   A 1.0000 MHz TTL compatible digital oscillator.
(5)   Nine 14-pin wire wrap sockets.
(6)   Three modular phone jacks, one each for the counter, timer and voltage measurer (as described in Section 8.2).
(7)   Telephone cables to connect the jacks to experimental apparatus.
(8)   A 40-conductor socket and ribbon cable to connect the CDI and CMS modules (as described in Section 7.7.1).
(9)   A 16-conductor socket and ribbon cable to connect the CMS and ADC modules (as described in Section 8.2).
(10) A wire wrap card identical to the CDI's (as described in Section 7.7.1).
(11) A bag of individual wire wrap terminals.
(12) A wire wrap tool and several spools of different color wire.
(13) Four or five .01 μf disc capacitors.
        Vendors for the components are in Appendix B.

9.7.2 Wire Wrap Card Layout

Figure 9.19 gives a layout for the CMS wire wrap card. If there is a ground plane, it should be on the top or components side. Solder and/or epoxy the twelve IC and two connector sockets approximately as shown. Epoxy the three phone jacks. Solder four wire wrap terminals in front of each jack as shown. Solder from left to right the yellow, green, red and black phone wires to the tops of terminals. Finally, solder four pairs of terminals up the left side of the card approximately as shown. Solder .01 μf disc capacitors between the terminals. If the card has a ground plane, solder the ground wire wrap terminals to the plane.

The three PIT's are labeled A, B and C. Similarly, the three 74LS74's are labeled A, B and C. The two 74LS04's are labeled A and B as are the two 74LS08's. A 7474 IC contains two flip-flops numbered 1 and 2 and the 7404 contains six INVERTERS numbered 1 to 6. Gate numbering for these and other IC's is in Appendix C. The gate labels in Figure 9.16 refer to the IC (letter) and gate in the IC (number).

9.7.3 Wire +5, Ground, BCK, S/S*, d_7...d_0 and the PIT Controls

The following sections are a guide for wire wrapping the CMS circuit. The drawings follow the style introduced in Chapter 7 and are useful for both construction and testing.

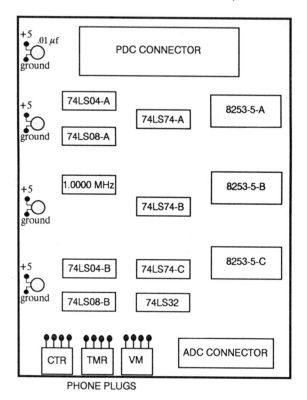

Figure 9.19. CMS circuit layout on a wire wrap card.

Figure 9.20 is the first drawing. The view is from the top, or components, side. Refer to the actual circuit and make the following connections.

(1) Wire +5 from the top wire wrap terminal to **H** on the CDI connector.
     Wire ground to both **L**'s on the connector.
(2) Wire +5 from the next lower wire wrap terminal to the **H**'s on 74LS08-
     A, 74LS04-A, 74LS74-A and PIT-A. Wire ground to the **L**'s on the
     same chips (avoiding loops).
(3) Wire +5 from the third wire wrap terminal to the **H**'s on the digital
     clock, 74LS74-B and PIT-B. Do likewise for ground to the **L**'s.
(4) Finally, wire +5 from the bottom terminal to all **H**'s on the remaining
     IC's, phone jacks and ADC connector. Do the same for ground to the
     L's.
(5) Next, wire the data lines (designated **0...7**) from the CDI connector to
     the three PIT's and ADC connector.
(6) Wire the PIT controls from the CDI connector to the three PIT's (wire
     all pins labeled **A**, all **B**'s, **C**'s and **D**'s).

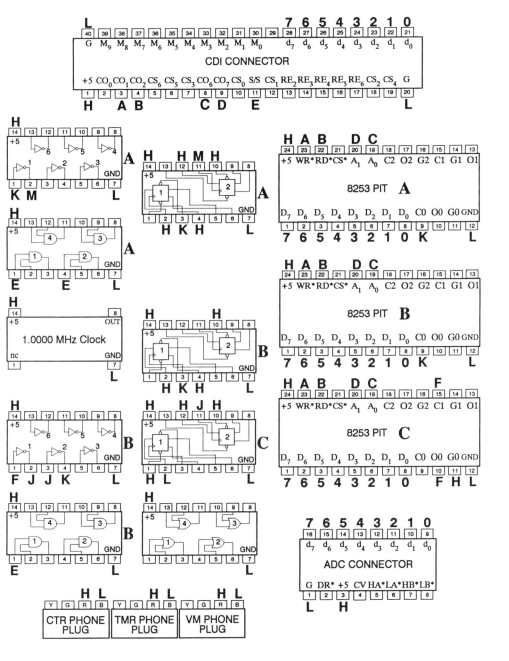

Figure 9.20. CMS wire wrap drawing showing +5, ground, S-S/S*, BCK, $d_7...d_0$ and PIT control connections.

(7)   S/S* (CDI pin 11) goes to AND gates A1, A2 and B1. Wire pins labeled **E**.

(8)   BCK originates with CTR_{C0} and goes to the clock of CTR_{C2} and to INVERTER-B1. Wire pins labeled **F**.

(9)   BCK* from B1 goes to FF-C2's clock input and to INVERTER-B2. Wire pins labeled **J**.

(10) BCK goes from B2 goes to FF-B1, PIT-B, FF-A1, INVERTER-A1 and PIT-A. Wire pins labeled **K**.

(11) Finally, wire BCK* from INVERTER-A1 to FF-A2 (**M**).

All that's left is circuitry specific to the counter, timer and voltage measurer.

9.7.4 Wire the Counter

Figure 9.21 is a guide to wiring and testing counter circuitry. Proceed as follows.

(1)   Wire the CDI's CS_6* to PIT-A's CS* (**A** to **A**).

(2)   Wire the EVENTS signal from the phone jack (yellow wire) to AND-A3 (**B**).

(3)   Wire the counter's EX-S/S* from the phone jack (green wire) to AND-A1 (**C**).

(4)   AND A1's output goes to the FF-A1, PIT-A's G_0, and the CDI's M_7. Wire all pins labeled **D**.

(5)   Wire Flip-flop A1's Q output to FF-A2's CLEAR* (**E**).

(6)   Wire FF-A2's Q output to AND's A3 and A4 (**F**).

(7)   Wire AND-A3's out to CK_1 and CK_2 (**J**).

(8)   Wire OUT_0 (from PIT-A) to AND-A4 (**K**).

(9)   AND-A4's out goes to G_1, INVERTER-A3 and the CDI's M_0. Wire all **M**'s.

(10) Finally, wire INVERTER-A3's output to G_2 and M_1 (**N**).
      This completes the counter.

9.7.5 Wire the Timer

Figure 9.21 is also a guide to wiring and testing the timer. Proceed as follows.

(1)   Wire the CDI's CS_5* to PIT-B's CS* (**0**).

(2)   Wire the timer EVENTS signal from the phone jack (yellow wire) to FF-B2's clock (**1**).

(3)   Wire the counter's EX-S/S* from the phone jack (green wire) to AND A2 (**2**).

(4)   AND-A2's out goes to FF-B2, PIT-B's G_0 and the CDI's M_8. Wire all **3**'s.

(5)   Wire FF-B2's Q output to FF-B2's DATA, G_1 and M_2 (**4**).

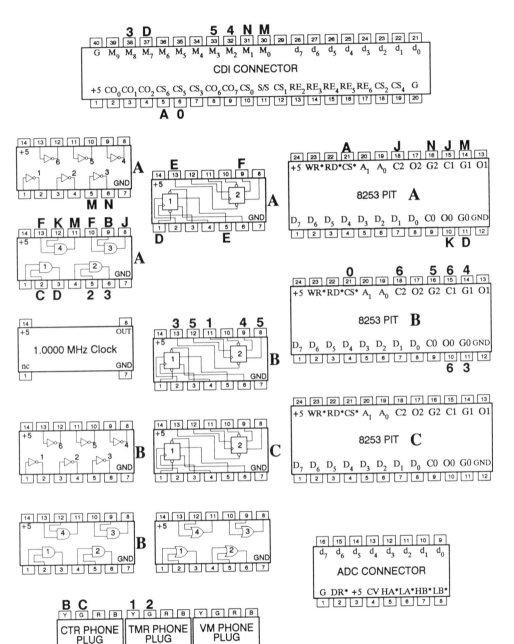

Figure 9.21. CMS wire wrap drawing showing counter and timer connections with letters and numbers, respectively.

(6)   Wire FF-B2's Q* output to G_2 and M_3 (**5**).
(7)   Finally, wire PIT-B's OUT_0 goes to CK_1 and CK_2 (**6**).
      This completes the timer.

9.7.6 Wire the Voltage Measurer

Figure 9.22 is a guide to wiring and testing the voltage measurer. Proceed as follows.

(1)   Wire the CDI's CS_3* to PIT-C's CS* (**A**).
(2)   Wire EX-TR from the voltage measurer's phone jack (yellow wire) to AND- B2 (**B**).
(3)   Wire the voltage EX-S/S* from the phone jack (green wire) to AND's B1 and B2 (**C**).
(4)   AND-B1's output goes to FF-B1, CDI's M_9 and PIT-C's G_1. Wire all **D**'s.
(5)   Wire AND-B2's output to AND-B3 (**E**). Wire AND-B3's output to the OR (**F**). Wire AND-B4's output to the OR (**G**). Wire the OR's output to G_2 (**J**). Finally, wire FF-C2's output to AND-B4 (**M**).
(6)   Wire the 1.0000 clock to CK_0 and CK_2 (**N**).
(7)   Wire OUT_2 to INVERTER-B6 (**P**). Wire B6's output to ADC connector pin 4 (**Q**).
(8)   Wire the ADC connector's pin 2 (DR*) to INVERTER-B4 (**R**). Wire B4's output to FF-C1's clock (**R**).
(9)   The CDI's RE_2* goes to FF-C1's SET* and to the CDI connector's HA*. Wire pins labeled **2**.
(10) Finally, wire the CDI's RE_3*, RE_4* and RE_5* to the ADC's LA*, HB* and LB*, respectively (**3** to **3**, **4** to **4** and **5** to **5**).

This completes the voltage measurer and CMS circuits. Visually inspect all work. Install the IC's.

9.8 Testing and Troubleshooting

In order to test the CMS circuit and software, assemble the system as follows:
(1)   Connect the parallel ports to the CDI module with appropriate cable(s).
(2)   Connect the CDI to the CMS module with a 40-conductor ribbon cable.
(3)   Connect the CDI to the ADC module with a 16-conductor ribbon cable.
(4)   Connect the power supply to the ADC module with a phone cable.

Before beginning, it's essential that the parallel ports, CDI and ADC modules work. If necessary, repeat the test procedures in earlier chapters. Troubleshooting tools are a logic probe, a set of IC test clips, a multimeter, a variable frequency digital clock and an oscilloscope.

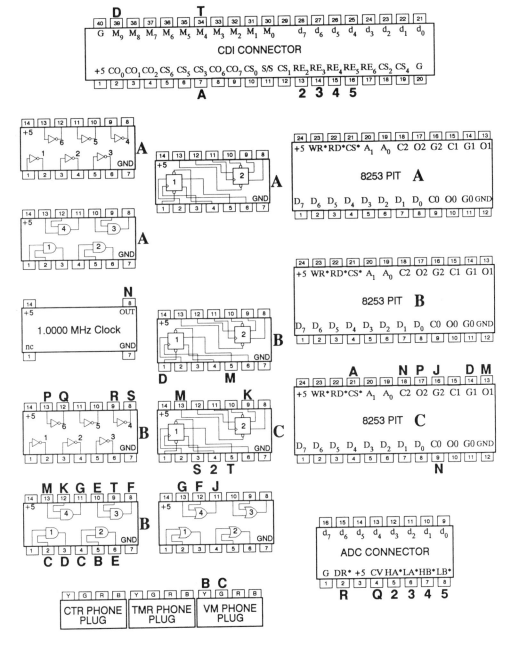

Figure 9.22. CMS wire wrap drawing showing voltage measurer connections.

9.8.1 Continuity Tests

With the ADC and CDI modules temporarily disconnected, test the CMS wiring by confirming that the resistance between pins which should be connected is zero. Put multimeter probes directly on IC pins and into connector sockets. Use Figures 9.20-9.22 as a guide and follow the order used in construction. The most common error is omitting a connection.

9.8.2 Test +5 and Ground

Connect the ADC module and turn the power on. Use the multimeter to measure +5 volts. If the value isn't at least 4, disconnect the CMS and measure +5 in the ADC module. If correct there, remove the IC's from the CMS, reconnect the ADC and measure again. If the voltage is still incorrect, look for a solder bridge or bad power supply noise capacitor. If correct after removing the IC's, systematically reinstall the chips to identify the bad one.

Check all +5 and ground connections in Figure 9.20 with a logic probe.

9.8.3 Test the Base Clock

The first operational test is the base clock. Load the software shell from Chapter 7 and set up CMS.BAS as described in Section 9.2.3. Print out the program and carefully proof. Then enter the following main program:

```
MAIN:  CALL PROGPIT(3,&H36)
       CALL LOADPIT (3,0,232,3)
       END
```

After execution, observe BCK on an oscilloscope. Since the loaded value $(256*3 + 232)$ is 1000 and the clock input to CTR_{C0} is 1.0000 MHz, BCK should be 1000.0 Hz. Verify the frequency.

If BCK is wrong, use the scope to check that CK_0 is 1.00000 MHz and G_0 is high. If CK_0 and G_0 are correct, put the CALL PROGPIT statement alone in an infinite loop and check PIT-C's CS*, WR*, A_0 and A_1. If they carry out the timing pattern in Chapter 6, look at $d_7...d_0$ when WR* is active and see if the value is 00110110 (36 hex). If it isn't, look for an incorrect data line. Put LOADPIT alone in an infinite loop and again look at the control and data lines. If everything checks out but BCK is still wrong, replace the 8253.

Recall that an infinite QuickBASIC loop is exited by typing CTRL-BREAK.

The objective of these tests is to find an incorrect signal and then systematically look for the cause. Because wire wrap connections are tedious to undo, it's difficult to determine whether a problem is with the origination of a signal or the gates to which it is connected. As always with troubleshooting, patience and persistence as well as understanding the hardware and software are essential.

9.8.4 Test the Counter

Enter the counter test/demonstration program in Section 9.2.3. Print and proof. Wire EX-S/S* high. Be sure the timer and voltage measurer's EX-S/S's are low. Connect the digital clock set on 10,000 Hz to the EVENTS input. Run the program selecting the 0.1 second counting interval and 20 intervals. Data collection should take 20 * (.1) = 2 seconds and the number of counts per interval should be 1000. If expectations are fulfilled, try other clock frequencies and counting intervals, and test the external abort by taking EX-S/S* low. Then, proceed to the next section.

If the count values are incorrect, put the program in an infinite loop by temporarily inserting GOTO WCTR1 just before the data collection loop's NEXT I. Run the program selecting 0.01 second intervals (it doesn't matter how many). Check the following with either the logic probe or oscilloscope:
(1) C-S/S (M_7) should be high. If not, check EX-S/S* and S-S/S*.
(2) SYNC should also be high. If not, check the flip-flops and BCK.
(3) Use the oscilloscope to verify that BCK reaches PIT-A's CK_0.
(4) Use the scope to look at G_1 and its inverse G_2 as they reach CTR_1 and CTR_2. They should have a half period of .01 second. If not (and everything else so far is correct), the PIT may not be programmed correctly. Check the software including constants. Since PIT-C CTR_0 correctly originates BCK, the likely hardware problem is an improperly wired CS_6* or a defective 8253.
(5) If the program and load operations appear correct but the count values are still wrong (after returning to the original program), the problem may be the wait or READPIT software. The former is partially checked by choosing a long counting interval and many intervals and then seeing if acquisition takes the expected time (NINT * CI).
(6) Finally, put READPIT alone in an infinite loop and check the control signals.
If the digital clock which supplies the EVENTS input is "noisy" or is not "completely" TTL compatible, it's possible to get more than the expected number of counts. If this is the case, try other digital clocks including the 1.0000 MHz oscillator.

9.8.5 Test the Timer

After BCK and the counter check out, hardware problems with the timer and voltage measurer probably do not involve PIT's and can be rapidly found.

Enter the timer test/demonstration program in Section 9.3.3. Print and proof. Wire EX-S/S high. Use the digital clock set on 10 Hz (.10 second period) for the EVENTS input. Run the program. Select range 1 and 10 events. Data acquisition should take .100 * 10 = 1.0 second and the values should be .100 second. If expectations are fulfilled, test other clock frequencies and ranges and then proceed to the next section.

If the values are incorrect, set up an infinite loop by inserting GOTO WCTR1 just before the data acquisition loop's NEXT I. Run the program and again select range 1. BCK should be 500 KHz and TCK should be 100 KHz. Use the logic probe or oscilloscope to check the following:

(1) Verify BCK's frequency as it reaches PIT-B's CTR_0.
(2) G_0 should be high. If not, check EX-S/S* and S/S*.
(3) OUT_0 (TCK) should be 100 KHz. If not and BCK and G_0 are correct, then PIT-B's CTR_0 is incorrectly programmed or loaded, or the chip is defective. Check the software including constants. Check CS_5*. Finally, replace the 8253.
(4) Next, look at G_1 and G_2 on the scope. They should have half periods of .100 second. If not, there is a problem with the flip-flop.
(5) If everything checks out but the values are still wrong (after returning to the original program), the problem is probably with the wait or READ-PIT software.

9.8.6 Test the Programmable Rate Voltage Measurer

Enter the voltage test/demonstration program in Section 9.4.5. Print and proof. Connect known DC voltages (perhaps from batteries) to VA and VB in the ADC module. Wire EX-TR low and EX-S/S* high. Run the program and select 100 measurements/second and 20 total measurements. Data acquisition should take 20*.01 = .20 second. VA and VB values should equal the voltage of their respective sources. If expectation are met, test other cases and then proceed to the next section.

If the values are wrong, add GOTO WFLOW just before NEXT I in the data acquisition loop. Run the program and again select a rate of 100 (the number of measurements doesn't matter). Use an oscilloscope or logic probe to check the following.

(1) BCK should be 5000 Hz. Verify with the scope.
(2) VCK (at OUT_1) should be 100 Hz. If it isn't, check the software including constants.
(3) Check that VCK reaches G_2. If not, check with the logic probe that EX-TR at the OR gate is low and that SYNC is high.
(4) Connect OUT_2 (CV*) to oscilloscope channel 1 and CV as it reaches the ADC's to channel 2. They should be 1 μsec pulses. Then, connect channel 2 to DR as it reaches flip-flop C1. It should go low shortly after CV* goes high and return high 20 μsec later. If CV* or DR are wrong, check the inverters, ADC cable and software for PIT-C CTR_2. It's pos-

sible that with a 100 Hz measurement rate the scope can't "see" CV* and DR. If necessary, remove all read device calls except 2, all calculations and the INKEY statement from the data acquisition loop and then run the program at 500 or 1000 measurements/second.

(5) When CV* and DR check out, leave CV* on channel 1 and connect VM-DR* (FF-C1's output) to channel 2. It should go low shortly after CV* and return high when software reads the data. If not, put RE_2* on channel 2. If it doesn't go low sometime after CV*, check the software and then the flip-flop.

(6) When FF-C1 works but the voltage values are still wrong (after returning to the original program), the problem is with the read device software, data computations or defective ADC's.

9.8.7 Test Voltage Trigger

Enter the voltage analyzer program in Section 9.4.6. Print and proof. Wire EX-S/S* high. Connect EX-TR to the digital clock set on 10 Hz. Supply a DC voltage to VA's input (in the ADC module). Run the program for 10 seconds. There should be 100 "counts" in the array element (or adjacent array elements) corresponding to the DC voltage. If expectations are met, try other EX-TR frequencies.

If the values are wrong, use the oscilloscope to check that EX-TR is correctly reaching G_2. If not, look at VM-DR* as it reaches AND-B3. If programmable rate measurement works, the likely problem here is with software.

9.8.8 Determine the "Maximums"

In Sections 9.3, 9.4 and 9.5, methods were suggested to determine the minimum counting interval, the shortest time between events and the maximum rate of voltage measurement. After the CMS hardware and software check out, determine these quantities.

9.9 Conclusions

The CMS applications module measures counts, times and voltages, and serves a variety of common data acquisition needs. However, situations are certain to arise which the CMS cannot accommodate, as illustrated by several of the following exercises. An advantage of the parallel approach described in this book is the relative ease of creating new applications modules which satisfy the situations.

The most significant CMS limitation is the "speed" of measurement. In all cases, software must read, process and store voltage and count values as they are acquired. The next chapter shows how to overcome this limitation by putting data directly in memory.

Exercises

1. What base clock (BCK) frequency occurs if CTR_{C0}'s HIGH and LOW load bytes are:

   HIGH = 62, LOW = 27
   HIGH = 0, LOW = 11

2. If the digital clock in Figure 9.2 were 2.0000 MHz, what LOW and HIGH load bytes are needed for the following BCK frequencies:

   1.0000 MHz, 250 Hz, 3000 Hz

   If the exact frequency is not attainable, solve for the closest possible case.

3. Consider a separate clock generator applications module connected to the CDI. The goal is to simultaneously produce and make available to outside circuits 10 individually programmable clock signals. Each clock's frequency is selected from the range 10 KHz to .4 Hz. After initiation, the frequencies are maintained until "any key" is typed, whereupon one or more may be changed or the program terminated. Several 8253's are required. One counter should act as a BCK. Draw the circuit in the format of Figure 9.2. Start with the software shell in Chapter 7. Write a main program and list constants which are not in the shell. Careful planning and organization are required.

4. Consider a separate clock generator applications module connected to the CDI. The single output has one of these frequencies: 1.0 MHz, 100 KHz, 10 KHz, 1000 Hz, 100 Hz, 10 Hz, 1.0 Hz or .10 Hz. Digital switches (or some other external circuitry) select which frequency goes to the output. However, when the switches supply a designated value, software inputs and then controls which frequency goes to the output. Draw the circuit in the format of Figure 9.2. Start with the software shell in Chapter 7. Write a main program and list constants which are not in the shell. Don't forget the multiplexer presented in Section 2.9.

5. For the CMS counter in Section 9.2, what is the maximum range of "reasonable" count intervals (CI's) if the base clock (BCK) frequency is:

20 Hz, 3333.3 Hz, 200 KHz

Assume the minimum possible CI is 1.0 ms.

6. Determine reasonable CTR_{C0} and CTR_{A0} load values to achieve the following CI's for the CMS counter:

    50 sec, .080 sec, .025 sec

    If the CI cannot be exactly produced, do the closest possible case.

7. What's the worst consequence of leaving the SYNC flip-flops out of the CMS counter? Answer for the specific case of CI = .100 second, NINT = 10 and the EVENTS input connected to a 1000 Hz digital clock.

8. When the CMS counter's test/demonstration program runs, what values are displayed if:

    CI = .01 sec, NINT = 10 and EVENTS = 1.0000 MHz clock.
    CI = .80 sec, NINT = 1, and EVENTS = 1250 Hz clock.
    CI = .80 sec, NINT = 4, and EVENTS = .25 Hz clock.
    CI = 20 sec, NINT = 8, and EVENTS = 2800 Hz clock.

9. Write a CMS counter program which makes CI = 10 seconds, acquires data for 200 seconds, and afterward displays the number of times the count exceeded 500. Start with CMS.BAS in Section 9.2.3. Give only a main program and list the new and/or different constants.

10. Write a CMS counter program for the following situation. A signal is connected to EX-S/S* which occasionally goes high for up to 10 seconds during which a separate signal, connected to the EVENTS input, contains between 1000 and 25,000 events per second. The goal is a program which acquires data every time EX-S/S* is high and afterward computes and displays the average number of events per second while EX-S/S* was high. Assume there are at least several tenths of a second between EX-S/S* highs. Start with CMS.BAS in Section 9.2.3. Give only a main program and list the new and/or different constants.

11. Design a separate counter applications module connected to the CDI which analyzes real time changes in an extremely high frequency digital signal. To deal with the high frequencies, CI must be 10 μsec which is too short for the CMS counter's software. Devise a hardware scheme in which one PIT counter produces CI and a second produces a delay pulse 10 ms long. The end of CI triggers the delay and the end of the delay triggers the next CI and so on. Software reads data during the delay. The process repeats a selected number of times after which the frequencies and elapsed times are computed and displayed. Draw the

circuit in the format of Figure 9.2. Start with the software shell in Chapter 7 and give a main program and list the new and/or different constants. Don't forget about PIT mode 1.

12. Design a new applications module connected to the CDI with counters for 10 EVENTS signals. All simultaneously count for a CI selected from some reasonable range. To implement the module with the CMS scheme requires 20 counters (10 to count while the other 10 are read and vice versa). A different approach is to have one PIT counter produce CI and another produce a "known" delay which is long enough for the 10 counters to be read. The end of CI triggers delay and the end of delay triggers the next CI and so on. Draw the circuit in the format of Figure 9.2. Write a program which inputs both CI and the delay (selected from a reasonable list) and then inputs the number of counting intervals. After acquisition, values from all 10 counters and elapsed time are displayed. Start with the software shell in Chapter 7. Give a main program and list the new and/or different constants.

13. A problem discussed in Section 6.10 was the necessity for at least one event during all CI's. Several solutions were suggested for the no events possibility. Design a new applications module connected to the CDI which is identical to the CMS counter except an additional 8253 generates one event at the beginning of each CI. The signal (perhaps 1 or 2 μsec long) must go high after CI and must be combined with the actual EVENTS in some way. For simplicity, assume the actual signal is low when there are no events. Draw the circuit in the format of Figure 9.2. Start with the software shell in Chapter 7. Give a main program and list the new and/or different constants.

14. For the CMS timer in Section 9.3, determine the maximum range of times which can be measured with 1% accuracy when the following values are loaded in CTR_{C0} and CTR_{B0}:

     C0 = 1000, B0 = 100
     C0 = 2, B0 = 10
     C0 = 33, B0 = 33

     Assume the minimum time interval is 1 ms.

15. For the CMS timer in Section 9.3, determine reasonable CTR_{C0} (BCK) and CTR_{B0} (TCK) load values to achieve the following 1% accurate time measurement ranges:

     1.0 to 750 seconds
     .15 to 1.5 seconds

16. Sketch how the CMS timer circuit in Section 9.3 can be modified to respond to every other event. In other words, G_1 and G_2 alternate on every second event (as opposed to every event).

17. When the CMS timer's test/demonstration program runs, what values are displayed if TCK is 10 Hz, NEVT is 8 and the EVENTS signal is a clock with a period of 1 hour? How long does data acquisition take?

18. Write a CMS timer program which determines the time between every 10 events for 100 seconds. After acquisition, the program displays the number of 10 event sequences and the time of each. Assume events are always between 100 and 200 ms apart. Start with CMS.BAS in Section 9.2.3. Give only a main program and the new and/or different constants.

19. The CMS timer can determine frequency by measuring period and inverting the result. Write a program which repeatedly displays the frequency of a signal applied to the EVENTS input. Assume the range is 100 Hz (.01 second period) to .01 Hz (100 second period). Follow a reprogramming scheme similar to the counter's frequency meter in Section 9.2.5. Start with CMS.BAS. Give the main program and list the new and/or different constants.

20. Design a separate timer applications module connected to the CDI which has two EVENTS inputs and two TCK's, but otherwise works the same as the CMS timer. Write a program which determines the times between separate successions of events and after acquisition displays the times and elapsed time. Use a scheme similar to the CMS timer to select ranges. Since data might be ready from both timers at once, the minimum timing interval must be extended. Draw the circuit in the format of Figure 9.2. Start with the software shell in Chapter 7. Give a main program and list the new and/or different constants.

21. The CMS timer measures the interval between when the signal on the EVENTS lines goes high and the next time it goes high. Suppose it's important to measure the length of high pulses as shown in Figure 9.23 and still keep track of elapsed time. Design a new applications module connected to the CDI which does this. Specifically, one PIT counter measures time when the signal is high and another when it is low. Write a test/demonstration program which operates the same as the CMS timer. Draw the circuit in the format of Figure 9.2. Start with the software shell in Chapter 7. Give a main program and list new and/or different constants.

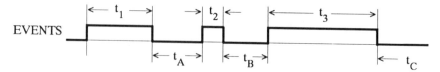

Figure 9.23. Measurement specifications for Exercise 21.

22. For the CMS voltage measurer in Section 9.4, what conversion rates occur if CTR_{C0} (BCK) and CTR_{C1} (VCK) are loaded with the following values:

   $C0 = 2$, $B0 = 20,000$
   $C0 = 25,000$, $B0 = 25,000$

23. For the CMS voltage measurer, what CTR_{C0} (BCK) and CTR_{C1} (VCK) load values produce the following conversion rates:

   850 Hz, 12 Hz, once an hour

   If the exact rate cannot be achieved, do the closest case.

24. Suppose the CMS voltage measurer's EX-TR is wired low, VA is a digital clock with a frequency of 200 Hz, and VB is 5.00 volts DC. Describe the displayed results when the test/demonstration program runs with RATE equal 500 and NVR equal 50.

25. Suppose the CMS voltage measurer's EX-TR is a 10 Hz digital signal and VA is a periodic 0 to 8 volt triangular wave with a .10 Hz frequency. What results are displayed after the voltage analyzer program in Section 9.4.6 collects data for 1 hour?

26. Repeat Exercise 25 for the case of EX-TR = 100 Hz, VA connected to a 0 to +5 volts square wave of frequency 10 Hz and data collection lasting 100 seconds.

27. Write a program for the CMS voltage measurer which determines as fast as "safely" possible 10 values at 10-bit accuracy from VA and afterward displays the average voltage. The sequence repeats every time "any key" is typed. Start with CMS.BAS in Section 9.2.3 and give a main program and list the new and/or different constants.

28. Suppose EX-S/S* for the CMS voltage measurer is a symmetric 1.0 Hz digital clock. Write a program which determines voltage at 8-bit accuracy from VA while EX-S/S* is high. Repeat the process for 10 EX-S/S* cycles and keep the data for each sequence in separate arrays (or perhaps in a multi-dimensional array). Set VCK to 200 Hz. After ac-

quisition, display the average voltage for each cycle. Start with CMS.BAS in Section 9.2.3. Give a main program and list the new and/or different constants.

29. Write a voltage analyzer program, similar to the one in Section 9.4.6, which uses all 10 bits from both VA and VB. Instead of the external trigger, use the programmable rate scheme to convert 10 times a second. Start with the voltage test/demonstration program and give a main program and list the new and/or different constants.

30. Suppose it's necessary to measure two voltages at different rates. Design a new applications module connected to the CDI which accomplishes the task. The first thing to realize is that a new ADC module is necessary so that ADC-A and ADC-B respond to their own CV's and send their own DR*'s. Each must also have its own VCK which triggers CV. The ADC cable and sockets must be expanded to at least 20 lines. Elapsed time for each voltage is determined by multiplying its conversion rate by the number of readings. Draw the application circuit in the format of Figure 9.15. Draw the new ADC module in the format of Figure 8.1 which includes cable socket pin assignments. Write a program which acquires a selected number of voltages from both ADC's. Input the measurement rates for both VCK's in a way similar to the voltage test/demonstration program. Care must be taken to be sure both rates can be successfully monitored and data read when ready. Start with the software shell in Chapter 7. Give the main program and list the new and/or different constants.

31. Suppose it's necessary to simultaneously measure 10 separate voltages at a programmable rate. Design a new applications module connected to the CDI which accomplishes the task. The first thing to realize is that a new ADC module with 10 ADC's is necessary. (A multiplexed ADC could also be considered.) The same CV goes to all ADC's and only one DR* is needed. However, 10 HE*'s and 10 LE*'s are required. The ADC cable and sockets must be expanded to 32 lines. The application circuit is identical to the CMS except for the expanded HE*'s and LE*'s. A problem is the insufficient number of RE*'s. However, unneeded CS*'s could be used if new constants are developed for the READD subroutine. Draw the new module in the format of Figure 9.15. Draw the new ADC module in the format of Figure 8.1 which includes cable socket pin assignments. Write a program identical to the test/demonstration except the number of voltage sources and accuracy are input parameters. Care must be taken to be sure the selected rate, accuracy and number of voltages can be successfully accommodated.

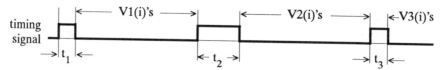

Figure 9.24. Timing pulses for Exercise 32.

Start with the test/demonstration program in Section 9.4.4. Give the main program and list the new and/or different constants.

32. Suppose it's necessary to do the combination of time and voltage measurements shown in Figure 9.24. Specifically, time is measured when the signal is high and voltage measured when it is low. The sequence starts when a separate hardware S/S* signal goes high and continues until it returns low. The EX-S/S* of the CMS voltage measurer and the EVENTS signal of the timer cannot accommodate this case. Design a new CMS module connected to the CDI which does. Draw the circuit in the format of Figure 9.15. Write a program which performs the measurements. Assume the high and low intervals of the control signal are never less than 100 ms. Read only one voltage at 8-bit accuracy. Keep time and voltage values in arrays. After acquisition, display the voltages, times and elapsed times. Start with the software shell in Chapter 7. Give the main program and list new and/or different constants.

33. Construct and test the CMS circuit. Use guidelines in Sections 9.7 and 9.8 as desired.

34. Enter the three test/demonstration programs in Sections 9.2.3, 9.3.3 and 9.4.4. Test each for cases where results can be anticipated.

35. Use suggested procedures in Sections 9.2.3, 9.3.3 and 9.4.4 to determine the minimum counting interval, the minimum time between events, and the maximum voltage measurement rate.

36. Enter and test the frequency meter program in Section 9.2.5. See how fast the program keeps up with "quick" changes.

37. Enter and test the voltage analyzer program in Section 9.4.6. Carry out Exercise 25. It's likely that the anticipated results will not be found. There are subtle relationships between the sampling rate and frequencies in the Fourier series which compose the triangle.

38. Enter the time/voltage measurement program in Section 9.4.6. Use a digital clock for the EVENTS signal and a battery for the voltage. Set the clock on 10 Hz and see if the anticipated results occur.

10   Fast Voltage Measurer

10.1 Introduction

The last several chapters of this book describe a detailed approach to data acquisition using the I/O ports presented in Chapter 3. The parallel data collector, as it's called, consists of the ports, the control/data interface (Chapter 7), the ADC module (Chapter 8) and an applications module. Any computer with suitable ports can operate the hardware and support the software.

Chapter 9 presented an example applications module which determines counts, times and voltages. A characteristic of the measurements is that software waits for data ready and then inputs and stores values. Because of the time required for these operations, the minimum counting interval, shortest measurable time, and maximum conversion rate depend on the computer's speed and the software's efficiency. Cases may arise which require smaller intervals, shorter times and/or higher rates. This chapter presents an applications module which automatically converts voltages and stores the values in its own memory. The system increases the maximum rate and works identically on original PC's, Apples II's and the latest computers. The approach also applies to counting and timing measurements.

The Fast Voltage Measurer (FVM) determines the voltage of one source at 8-bit accuracy with the rate and number of conversions programmable. The maximum 40K measurements/second arises from the AD573's 20 μsec conversion time, and the maximum number of measurements equals 32K, the storage capacity of four 6264 8K IC's. Data acquisition starts when both Software Start/Stop (S-S/S*) and EXternal Start (EX-S) become high. Software monitors for the end of data acquisition after which the stored values are read and processed as desired. The system, shown in block form in Figure 10.1, requires the CDI and ADC modules, four 6264 memory IC's, address generating counters and one 8253 PIT.

Sections 10.2-10.5 describe FVM hardware. Section 10.6 outlines software and presents a test/demonstration program. Section 10.7 gives an example measurement. And Sections 10.8 and 10.9 cover construction and testing.

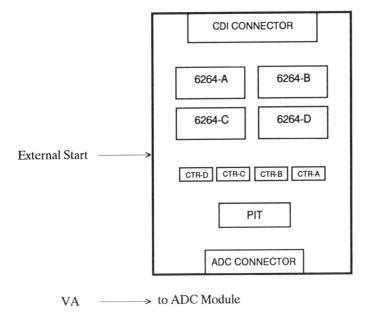

Figure 10.1. Block diagram of the Fast Voltage Measurer (FVM) applications module.

10.2 Memory and Address Generator Circuitry

Section 5.3 gave circuitry for a 6264 memory IC and address generating counters. The purpose of this section is to expand the circuit to four memory chips.

A 6264 has 13 address inputs ($A_{12}...A_0$) which allow access to 8192 memory locations. The address bits originate with four 74193 counters wired as shown in Figure 5.4. Starting at 0, the counter outputs undergo the following sequence.

memory location	$A_{12}...A_0$
first	0 0000 0000 0000
second	0 0000 0000 0001
third	0 0000 0000 0010
•••	•••
next to last	1 1111 1111 1110
last	1 1111 1111 1111

The four 6264's necessary for 32K bytes of memory are labeled A, B, C and D. Each chip's address inputs are connected to $A_{12}...A_0$. The problems are how to expand the address range and how to select which 6264 responds to read and write commands. A solution involves using the memory chip enables and A_{13} and A_{14} from the high address counter. The table below shows the binary addresses as they extend to the next higher 2 bits.

address	A_{14}	A_{13}	$A_{12}...A_0$	address range
first	0	0	0 0000 0000 0000	1st 8K, 6264-A
•••	•••	•••	•••	
8191	0	0	1 1111 1111 1111	
8192	0	1	0 0000 0000 0000	2nd 8K, 6264-B
•••	•••	•••	•••	
16483	0	1	1 1111 1111 1111	
16484	1	0	0 0000 0000 0000	3rd 8K, 6264-C
•••	•••	•••	•••	
24675	1	0	1 1111 1111 1111	
24676	1	1	0 0000 0000 0000	4th 8K, 6264-D
•••	•••	•••	•••	
32867	1	1	1 1111 1111 1111	

$A_{12}...A_0$ goes through all possible values when $A_{14}A_{13}$ is 00, and again when 01, 10 and 11. A reasonable scheme is to enable 6264-A when $A_{14}A_{13}$ is 00, 6264-B when 01, 6264-C when 10 and 6264-D when 11.

Recall from Section 5.1 that a 6264 responds to read and write commands only when CE_1^* and CE_2 are both active. For example, 6264-A responds to the first 8K addresses if its CE_1^* and CE_2 are 0 and 1, respectively, when $A_{14}A_{13}$ is 00. The following table shows how A_{13} and A_{14} and their inverses A_{13}^* and A_{14}^* enable each 6264 to respond to a unique 8K memory range.

A_{14}	A_{13}	enables 6264	with these connections CE_1^*	CE_2
0	0	A	$A_{14}=0$	$A_{13}^*=1$
0	1	B	$A_{14}=0$	$A_{13}=1$
1	0	C	$A_{14}^*=0$	$A_{13}^*=1$
1	1	D	$A_{14}^*=0$	$A_{13}=1$

Figure 10.2 is the memory/address generator.

In the circuit, $A_{12}...A_0$ goes from the counters to all four 6264's. Each chip's D8...D1 is connected to the bi-directional data bus $d_7...d_0$ (which in later circuitry goes to the PIT and the CDI and ADC modules). Also, the four memory Write Enables (WE*'s) are connected as are the four Output Enables (OE*).

The memory and address generator is hereafter drawn and thought about as shown in Figure 10.3. In addition to $d_7...d_0$, the connections are:

(1) LD* (LoaD) makes all counter outputs 1.

(2) A CU (Count-Up) low-high causes the low counter to increment (with carries to higher counters as required). The first CU after an LD* makes all address bits zero.

(3) When WE* returns high, the value currently on the data lines is stored at the address on $A_{12}...A_0$ in the 6264 enabled by $A_{14}A_{13}$.

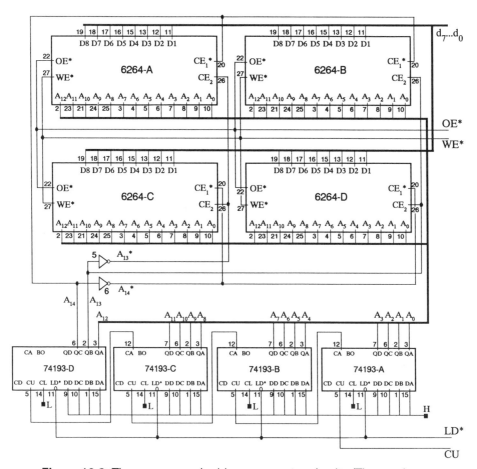

Figure 10.2. The memory and address generator circuit. (The numbers adjacent to the INVERTERS distinguish the gates in a 7404 IC and are used later.)

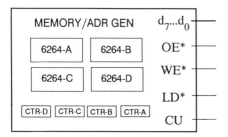

Figure 10.3. Block representation of the memory/address generator circuit in Figure 10.2.

(4)   An active OE* causes the 6264 enabled by $A_{14}A_{13}$ to put the value at the current address on the data lines.

10.3 Measurement and Storage Circuitry

After establishing the 32K memory and address generator, the next task is to devise a hardware scheme which repeatedly: (1) advances the address; (2) converts a voltage; and (3) stores the value at the address. The process is intricate and requires careful analysis of AD573 and 6264 operations.

10.3.1 Generation of Convert Pulses at a Programmable Rate

Two counters in an 8253 PIT generate convert pulses at a programmable rate. Figure 10.4 gives the circuit in the format developed in Chapter 9. The PIT is connected to the CDI's CS_0* and therefore responds to CALL PROGPIT(0,CW) and CALL LOADPIT(0,J,LOW,HIGH). CTR_0's clock input is a 1.0000 MHz digital oscillator. The gate is wired high (for now). When programmed for mode 3, OUT_0 is a clock whose frequency is 1,000,000 divided by the loaded value. Because of initiation considerations, the inverse of OUT_0 is VCK (the Voltage rate ClocK).

VCK supplies CTR_2's gate input. CK_2 is the 1.0000 MHz clock. OUT_2 is SVC* (Start a Voltage Conversion). When programmed for mode 1, each

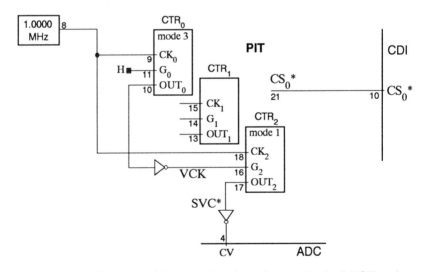

Figure 10.4. Circuitry which generates the voltage rate clock VCK and convert pulses. The CDI and ADC connectors are on the right and bottom, respectively. Not shown are PIT control lines (WR*, RD*, A_0, A_1) and CDI data lines $d_7...d_0$ to the PIT and the ADC module.

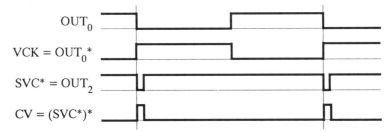

Figure 10.5. Timing diagram for the voltage rate clock and convert pulses as produced by PIT counters 0 and 2.

VCK low-high causes SVC* to go low for a length of time equal to the loaded value multiplied by CK_2's period $(1/1,000,000 = 1\,\mu\text{sec})$. With the load value 1, SVC* is a $1\,\mu\text{sec}$ low pulse. Since AD573-A requires an active high convert signal, CV is the inverse of SVC* and goes to the ADC connector. The timing pattern is in Figure 10.5.

CTR_1 controls the number of conversions as discussed in Section 10.4.

10.3.2 Control Flip-flop

Shortly after receiving CV, AD573-A takes its **D**ata **R**eady high. Until that time, the previous converted value is available. DR* returns low when conversion is complete and new data is ready. The tasks here are to work out when and how to advance the address and store converted values. A 7474 flip-flop plays the central role.

As shown in Figure 10.6, the flip-flop is set by activation of SVC* and cleared by the end of (DR*)*. The output SIM* (**S**tore **I**n **M**emory) goes to ADC-A's **H**igh byte enable (HA*) and to the memory's **W**rite **E**nable (WE*). In addition to making SIM* high, SVC* goes to the memory's **C**ount **U**p (CU).

The circuit carries out conversion and storage as shown in Figure 10.7 and summarized below.

(1) A sequence starts when OUT_0 goes low and its inverse VCK high. (This occurs at a rate determined by the value loaded in CTR_0.)

(2) VCK ($= G_2$) triggers SVC* ($= OUT_2$). The beginning of SVC* sets the flip-flop which makes SIM* high.

(3) The low-high end of SVC* ($1\,\mu\text{sec}$ later) increments the address.

(4) Shortly after the end of SVC*, DR* goes high and conversion starts.

(5) When conversion is finished, the ADC takes DR* low and (DR*)* clears the control flip-flop which makes SIM* low. A low SIM* causes the ADC to put the high 8 bits of the just converted value on $d_7...d_0$.

(6) Nothing happens until the next VCK low-high which again through SVC* sets the flip-flop (step 2 above). The action takes WE* ($= $ SIM*) high

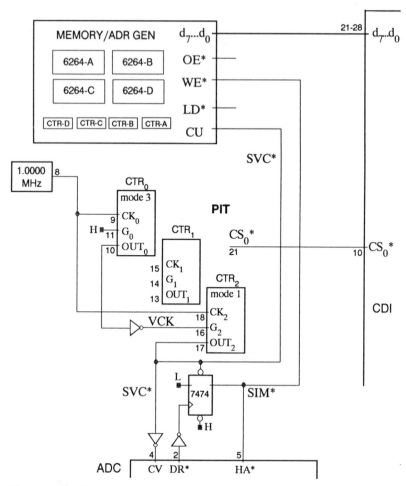

Figure 10.6. Circuitry which repeatedly advances the address, converts a voltage and stores the value in memory.

and the value converted just after the last VCK low-high is stored at the current address. Later, the address is advanced (step 3) and the next conversion starts (step 4). (When SIM* went high in step 2, the value from the previous sequence was stored.)

In essence, SVC*'s beginning stores the last converted value and its end increments the address and initiates the next conversion.

Since SIM = HA* = WE*, HA* and WE* go high simultaneously. Many devices which store values off data lines require the values to remain until after the "storing edge." However, 6264 specifications say WE* may go high at the same time data is taken off D8...D1, as shown in Figure 5.3.*

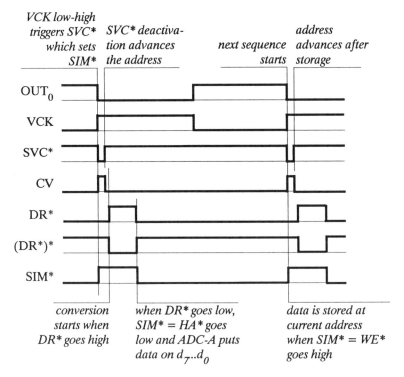

Figure 10.7. Timing diagram for hardware voltage conversion and storage. SIM* supplies the ADC's HA* and the 6264's WE*.

With the hardware conversion and storage scheme worked out, the remaining issues are initiation, termination, and software reading stored values.

10.4 Initiation and Termination Circuitry

10.4.1 Basic Plan

Fast voltage measurement is initiated either by software or hardware. The CDI's Software Start/Stop (S-S/S*) is ANDed with an EXternal Start (EX-S) to produce FV-S (Fast Voltage Start). The goal is for data acquisition to begin when FV-S goes high. If EX-S is wired high, software alone makes FV-S high. If EX-S is low when S-S/S* goes high, hardware holds up initiation. FV-S goes to PIT CTR_0's gate input (which was wired high in Figure 10.6). So, while FV-S is low, OUT_0 is high, VCK is low and no conversions occur. When FV-S goes high, G_0 goes high and CTR_0 starts producing VCK cycles.

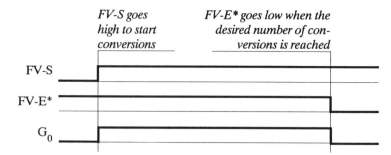

Figure 10.8. Timing diagram for initiation and termination of hardware voltage measurement. Conversions occur while G_0 is high.

Recall that when the gate to a mode 3 PIT is low, the output is forced high. For reasons which become clear with the final circuit and software, there is no external stop and software cannot terminate data acquisition by making S-S/S* low.

The more difficult task is terminating conversions after a programmed number. Suppose the new signal FV-E* (Fast Voltage End) is high until acquisition should end. FV-S is ANDed with FV-E* and the result replaces FV-S at G_0. Then, as shown in Figure 10.8, FV-S starts acquisition by making G_0 high and FV-E* stops acquisition by making G_0 low.

The next issue is how to produce FV-E*. Suppose unused PIT CTR_1 is programmed for mode 1, loaded with the desired number of conversions and wired to count VCK cycles. As shown in Figure 6.4, OUT_1 goes low after the first VCK ($= CK_1$) low-high-low and returns high when the count decrements to zero. So the signal's inverse, shown in Figure 10.9, is a possibility for FV-E*. However, there is a fatal problem. OUT_1* is low until VCK goes low-high-low which cannot happen until G_0 goes high. But G_0 can't go high until both FV-S AND FV-E* are high. Therefore, OUT_1* can't be FV-E*.

Additional circuitry might resolve the problem. But a simpler solution is to use PIT mode 0 which was not explicitly covered in Chapter 6. According to Intel's specifications, as soon as a counter is programmed for mode 0, the output is low. Then, with the gate high, the loaded value decrements every time the signal at the clock input goes high-low. When the count reaches zero, the output goes high. So an approach is to program CTR_1 for mode 0 and load the desired number of conversions. At this point, OUT_1 is low and its inverse OUT_1* = FV-E* is high. As shown earlier in Figure 10.7, OUT_0 goes high-low at the beginning of each conversion. So if OUT_0 is connected to CK_1 and G_1 is wired high, each conversion causes CTR_1 to decrement. When the count reaches zero, OUT_1 goes high and FV-E* low as shown in

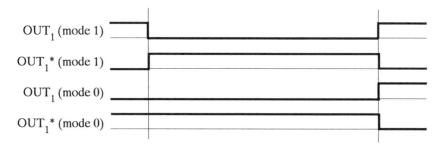

Figure 10.9. Operation of CTR₁ in mode 1 and in mode 0.

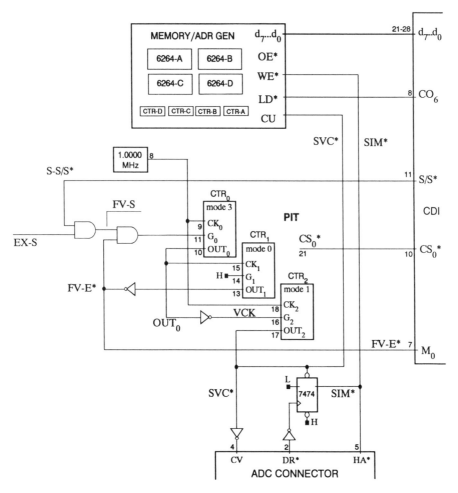

Figure 10.10. Initiation and termination circuitry. CTR_1 counts the number of conversions.

Figure 10.9. And a low FV-E* stops conversions as described above. Circuitry in Figure 10.10 accomplishes initiation and termination.

FV-E* is also connected to CDI monitor input M_0. After executing CALL START, software simply waits while $CI_0 = M_0$ is high.

10.4.2 Initiation Details

It's important to know several initiation details not emphasized in the basic plan. After software executes CALL SSTOP at the beginning of any main program, FV-S and G_0 are low. A low G_0 makes OUT_0 high and VCK low. A low VCK makes SVC* high. However, the state of the flip-flop and therefore SIM* is unknown. Recall that activation of SVC* sets the flip-flop and the end of DR* clears it and neither occur up to this point.

After the gate to a PIT counter goes high, the next clock low-high-low transfers the value loaded by software in the storage register to the count register and only thereafter does the count register decrement on clock high-lows. Therefore, the first OUT_0 cycle is delayed up to $1\ 1/2\ \mu sec$ after G_0 goes high as shown in Figure 10.11.

Conversions are triggered by VCK low-high transitions at G_2. If OUT_0 rather than VCK were used for this purpose, the first conversion would occur at the end of CTR_0's first cycle, as also shown in Figure 10.11. By using VCK,

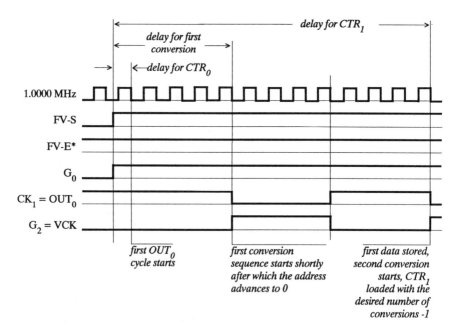

Figure 10.11. Initiation timing diagram.

the first conversion is after half a CTR_0 cycle (plus the OUT_0 delay). So, if VCK is 25 KHz, the net delay is $40\,\mu sec$ (plus up to $1\,1/2\,\mu sec$).

There may be occasions when this delay is unacceptable. However, more elaborate circuitry is required to produce the first conversion at the beginning of the first OUT_0 cycle.

The first OUT_0 low-high-low causes CTR_1 to transfer its loaded value from the storage to the **C**ount **R**egister (CR). Since the end of low-high-low coincides with storing the first voltage, CR at that point contains the **L**oaded **V**alue (LV). When voltage 2 is stored, CR = LV-1; when voltage 3 is stored, CR = LV-2. With this scheme, it's not until voltage LV + 1 is stored that CR = LV-LV = 0 and data acquisition terminates. Therefore, the loaded value should be one less than the desired number of conversions.

The timing diagram in Figure 10.7 shows a typical conversion sequence. As previously discussed, the end of each SVC* pulse advances the address. If LD* is activated by software before data acquisition starts, the first SVC* makes the address zero and that's where the first value is stored.

10.4.3 Termination Details

Several important termination details were omitted from the basic plan. When OUT_0 goes low and CTR_1 finally decrements to zero, FV-E* goes low. Coincident with OUT_0 going low, as always VCK goes high triggering SVC* which sets the flip-flop and makes SIM* high. When SIM* becomes high, data from the most recent conversion is stored. However, unlike previous sequences, FV-E* takes G_0 low, which after gate delays makes OUT_0 high and VCK low. Since VCK is connected to G_2, SVC* returns high after further gate delays. The process is shown in Figure 10.12.

Because the next step is to read stored values, the final states of signals connected to the memory/address generator are important. SVC* goes to CU and its final state is high. SIM* goes to WE* but it's final state is uncertain. When VCK goes high for the last time and then back low after gate delays, the resulting CV pulse may or may not be long enough to trigger a conversion. Consequently, the behavior of DR* is unpredictable. With the flip-flop's PRESET* high, DR* may or may not make SIM* low. The uncertainty is resolved in the next section.

10.5 Read Memory Circuitry

After a conversion sequence, software from Chapter 7 zeroes the address and enters a loop which reads a stored value and advances the address. For test purposes, it's also desirable for software to be able to write to memory.

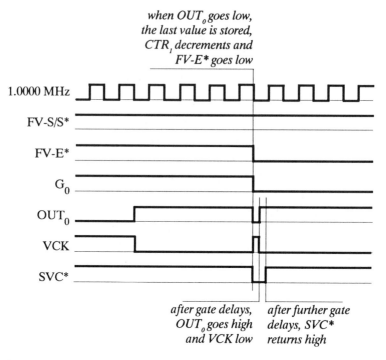

when OUT_0 goes low,
the last value is stored,
CTR_1 decrements and
FV-E* goes low

1.0000 MHz

FV-S/S*

FV-E*

G_0

OUT_0

VCK

SVC*

after gate delays,   after further gate
OUT_0 goes high   delays, SVC*
and VCK low   returns high

Figure 10.12. Termination timing diagram.

Therefore, the following CDI connections discussed in Chapter 5 and again in Section 7.3 must be made to the memory/address generator:

CDI signal	mem./adr. gen.
CO_1	WE*
CO_2	OE*
CO_6	LD*
CO_7	CU

Since LD* and OE* are not used for hardware storage, the issue is how to connect CU and WE* so both hardware and software can increment the address and store values.

The uncertain state of SIM* left by the termination sequence in the last section must be resolved. Since SIM* is connected to HA*, a low state would cause the ADC to leave a value on $d_7...d_0$ in conflict with read's use of the data lines. A solution is to AND FV-E* with SVC* to produce SVC′* which is connected to the flip-flop's PRESET* and to CU. Since FV-E* is definitely low after termination, the resulting low SVC′* sets the flip-flop and makes SIM* high. Even if the final CV triggers a conversion, the flip-flop doesn't respond to DR* since an active 7474 PRESET* overrules clock/data operations. On the other hand, while FV-E* is high, SCV'* equals SCV*. Therefore, the new AND gate does not interfere with the hardware measurement sequence as previously described.

With the additional AND, after termination of hardware measurement, the signals CU (SVC'*) and WE* (SIM*) are low and high, respectively. CO_7 (software count-up) is ORed with the SVC'* and the result connected to CU. Then, if either SVC'* or CO_7 is low, the low-high-lows of the other reach the address generator. CO_1 (software WE*) is ANDed with SIM* and the result connected to WE*. Then, if either CO_1 or SIM* is high, the high-low-highs of the other reach the memory. Therefore, after hardware data

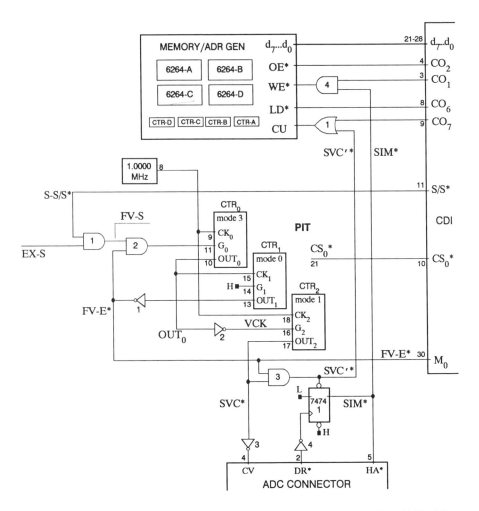

Figure 10.13. The complete fast voltage measurer circuit. The AND, OR, INVERTER and flip-flop gate numbers are used in Section 10.8.

acquisition, the memory subroutines in Chapters 5 and 7 work as previously described.

Figure 10.13 is the final circuit.

For software and testing considerations in the next sections, it's important to summarize the state of several controls just before and just after hardware measurement. The following table covers three cases.

(1) Case A is just before software starts measurement the first time the program is executed (after power is turned on). Stop has been called and the PIT counters programmed and loaded.

(2) Case B occurs after hardware measurement and CALL SSTOP but before software reads the stored values.

(3) Case C is the same as case A except data has been acquired since power was turned on.

signal	Case A	Case B	Case C
S-S/S*	L	L	L
FV-S	L	L	L
G_0	L	L	L
VCK	L	L	L
FV-E*	H	L	H
SVC'* (CU)	H	L	H
SIM* (WE*)	X	H	H

The flip-flop output SIM* is uncertain until after the first measurement.

10.6 Software

Software developed in this section carries out fast voltage measurement. The goal is a test/demonstration program which inputs a measurement rate and number of conversions, starts data acquisition, and afterward reads and displays the voltages. The starting point is the software shell in Chapter 7.

10.6.1 Fast Voltage Software Shell

The following CDI subroutines are needed: SSTOP, START, PROGPIT, LOADPIT, ZEROADR, ADVADR, READMEM, WRITEMEM and END-MEMWR. Actually, the latter two are not required but might be useful for test purposes. The unnecessary subroutines READPIT and READD are eliminated. Also, the array constants READPC(I,J) and REDD(I) are unnecessary and deleted. Since the PIT is connected to CS_0*, only program and load constants with index $I=0$ can be used and the others are eliminated. The modified shell is hereafter called FV.BAS.

Recall from the last section that CO_7 and SVC'* are ORed and the result goes to the address generator's CU. If CO_7 is high during hardware measurement, the OR output is high no matter what SVC'* is and hardware can't advance the address. So the last software operation before acquisition, CALL START, must be modified to be sure CO_7 is low. The original value of the constant STRT is C6 hex or 1100 0110 binary. If 0100 0110 or 46 hex is used instead, CO_7 goes low as soon as S-S/S* goes high. Also the last statement in the subroutine outputs NORM = &HFF. This must be changed to OUT CP,&H7F. The new constant and START routine are:

```
STRT = &H46

SUB START
    OUT CP, NORM
    OUT CP, STRT
    OUT CP, STRT
    OUT CP, &H7F
END SUB
```

It's not reasonable to change NORM to 7F hex since it's used by other subroutines.

An alternate approach is to not change STRT and the subroutine but simply to OUT CP,&H7F immediately after CALL START. However, there is a problem. As soon as START executes the first OUT CP,STRT, S-S/S goes high and acquisition begins. But CO_7 does not go low until after return. During this time, several conversions might occur but the address is not advanced.*

Finally, even with one PIT, it's still a good idea to use arrays for the load bytes, LOW(I,J) and HIGH(I,J), and for the control word, CW(I,J).

In all fast voltage software, three PIT counters must be programmed and loaded. The following table gives the mode, index, control word and low and high load bytes.

counter	mode	index	CW	LOW	HIGH	function
0	3	0,0	&H36	X	X	generate VCK
1	0	0,1	&H70	X	X	end conversions
2	1	0,2	&HB2	1	0	generate SVC*

"X" means the value is determined for specific applications programs. (Review Section 6.4.2 on control word formulation.)

The fast voltage shell, FV.BAS, follows. The subroutines are not repeated here but are the same as in Section 7.7.3 except for the modified START.

```
'       FV.BAS

DEFINT A-Z
DIM SELPP(8),PROGP(8)
DIM SELPC(8,3),LOADPC(8,3),READPC(8,3)
DIM LOW(7,3),HIGH(7,3),CW(7,3)
```

```
      COMMON SHARED CP,DP,NORM,STRT,SSTP
      COMMON SHARED SELPP(),PROGP(),SELPC()
      COMMON SHARED LOADPC(),READPC()
      COMMON SHARED MAADV,MALOAD,MREAD,MWRITE,MENDW
      CP = &H300 : DP = &H301
      NORM = &HFF
      STRT = &H46 : SSTP = &HCE
      CW(0,0) = &H36
      CW(0,1) = &H70
      CW(0,2) = &HB2 : LOW(0,2) = 1 : HIGH(0,2) = 0
      SELPP(0) = &HC7
      PROGP(0) = &HC5
      SELPC(0,0) = &H07 : SELPC(0,1) = &H47
      SELPC(0,2) = &H87
      LOADPC(0,0) = &H05 : LOADPC(0,1) = &H45
      LOADPC(0,2) = &H85
      MAADV = &H7F: MALOAD = &HBF
      MREAD = &HFB: MWRITE = &HFD: MENDW = &HC7
MAIN: CALL SSTOP
        . . .
      END
```

The memory subroutines follow including SSTOP and the modified START.

10.6.2 Test/demonstration Program

FV.BAS is the starting point for the test/demonstration program. The first operation after CALL SSTOP is to input the desired measurement rate and number of conversions.

The rate is input to the real variable VCK! (Voltage ClocK). The next task is to compute CTR_0's load value (LOAD0) from VCK!. Since VCK! = 1,000,000/LOAD0, LOAD0 = 1,000,000/VCK!. The AD573's 20 μsec conversion time limits the maximum rate as discussed in Section 10.6.3.

NCV is the selected Number of ConVersions. The range of possible values is 1 to 32K, the amount of memory available.

High and low load bytes for CTR_0 and CTR_1 are determined from VCK! and NCV as follows.

```
MAIN: CALL SSTOP
      CLS
      PRINT "Enter Freq in Hz, No. of Conversions"
      INPUT " --> "; VCK!, NCV
      LOAD0 = INT(1000000 / VCK!)
      HIGH(0,0) = INT(LOAD0 / 256)
      LOW(0,0) = INT(LOAD0 - (HIGH(0,0) * 256))
      HIGH(0,1) = INT((NCV - 1) / 256)
      LOW(0,1) = INT(NCV - 1 - (HIGH(0,1) * 256))
```

Recall from Section 10.4.2 that CTR_1 is loaded with the desired number of conversions - 1. HIGH(0,0), LOW(0,0), HIGH (0,1) and LOW (0,1) are computed so the load values equal 256*HIGH + LOW.

As discussed in Section 9.4.4, because LOAD0 must be an integer, "every" measurement rate is not possible and the gaps between available VCK's grow with higher frequencies. For example, suppose the input rate is 22,000 Hz. Software computes LOAD0 to be 45 and the actual programmed VCK is 22,222 Hz. A good idea is to display the actual measurement rate with the results.

The three counters are programmed and loaded as follows.

```
FOR J = 0 TO 2
  CALL PROGPIT(0,CW(0,J))
  CALL LOADPIT(0,J,LOW(0,J),HIGH(0,J))
NEXT J
```

The next three statements initiate hardware data acquisition. The modified START routine is called.

```
INPUT "Type Any Key to Start "; A$
CALL ZEROADR
CALL START
```

The subroutine ZEROADR first activates LD* (CO_6) which makes the address bits all 1, and then activates CU (CO_7) which advances the address to zero. As discussed at the end of Section 10.5, SVC'* is high before and after acquisition. Since SVC'* is ORed with CO_7, the software CU is not seen. Therefore, when ZEROADR is called, 1's are loaded into all four counters but the address is not advanced. Although unplanned, this is fortunate since the conversion sequence advances the address to zero before the first value is stored as shown in Figure 10.7.

Since M_0 (= FV-E*) starts high and goes low only when data is ready, the next software operation is to **Wait While M_0 is High**.

```
WWH:   IF (INP(CP) AND &H01) = 1 THEN GOTO WWH
```

The mask for M_0 is 0000 0001 or 1.

When hardware takes M_0 low, the next software operations are to send stop, zero the address, and read and display the stored values. The following statements accomplish these tasks.

```
CALL SSTOP
CALL ZEROADR
CONV! = 10/256
PRINT "Voltages from ";NCV;" Conversions"
PRINT "The Measurement Rate VCK is ";
PRINT USING "#####.##"; 1000000/LOAD0
FOR I = 1 TO NCV STEP 10
  FOR J = 0 TO 9
```

```
        CALL READMEM(SVOL)
        CALL ADVADR
        PRINT USING "##.##"; SVOL * CONV!;
        PRINT " ";
    NEXT J
    PRINT
NEXT I
```

The address must be zeroed before reading. Since FV-E* and therefore
SVC'* are low after acquisition, CU and ADVADR work properly. The
constant CONV! **CONV**erts the 8-bit binary voltages to actual values be-
tween 0 and 10. As discussed in Section 4.1 CONV! = 10/256. 1,000,000/
LOAD0 equals the actual programmed VCK!. SVOL is a **S**tored **VOL**tage.
Ten formatted values are displayed per row.

For the purposes of the test/demonstration program, the voltages are
not saved. They could be stored in an array, written to a file and/or graphed.

The final operation is a repeat option.

```
INPUT "Enter R to Repeat, Else End "; A$
IF A$ = "R" OR A$ = "r" THEN GOTO MAIN
END
```

10.6.3 Fastest Measurement Rate

The fastest voltage measurement rate is determined by the speed of the ADC.
The average conversion time of an AD573 is $20\,\mu$sec. Adding $5\,\mu$sec for the
convert pulse and activation of HA* and WE*, the maximum rate is approximately $1/(25\,\mu$sec$) = 40$ KHz.

The rate can be experimentally estimated. Enter the test/demonstra-
tion program. Connect a 1000 Hz digital clock to the voltage input. Run the
program at 25 KHz for 200 conversions. From the displayed results, count
the number of values in a cycle (the voltages will be $+5$ or so and 0 or so and
cycles are easy to see). There should be 25. Repeat at higher rates until the
expected number is not found. Use the actual VCK frequency as opposed to
the input rate.

10.6.4 Fast Voltage Capabilities and Limitations

The FVM's specifications are summarized below. Also, suggestions are made
on how capabilities might be expanded.
(1) Measurement rates are from 20 Hz to 40 KHz. The CMS system in
     Chapter 9 works for lower VCK's. Greater VCK's require faster ana-
     log-to-digital converters which could be incorporated in new ADC mod-
     ules.
(2) The maximum number of conversions is determined by the 32K bytes
     available in four 6264 memory IC's. Capacity could be expanded in a

new applications module with additional 6264's or other memory chips (such as the 62256 32K by 8-bit static RAM).

(3) Only one voltage can be measured. It's possible for a modified FVM to convert and store voltages from both AD573's, perhaps one after the other. Also, the ideas presented here could be applied to new ADC and applications modules where multiple voltages are simultaneously converted and stored in multiple sets of memories.

(4) Accuracy is 8 bits. All 10 AD573 bits could be used in a modified circuit by either adding memory or reducing the maximum number of conversions. Also, 12- or 16-bit ADC's are possibilities.

Exercises at the end of this chapter explore more powerful FVM's as well as fast counting and timing systems.

10.7 Example Measurement

The hardware presented in this chapter supports a variety of measurements. The most common case is voltage vs. elapsed time as carried out by the test/demonstration program. Another approach is to connect a digital clock to external start (EX-S) and program the system so every time the clock goes high, 100 conversions occur at a rate of, say, 40 KHz. The average voltage is computed and displayed and software then waits for the next EX-S high to repeat the sequence. A different concept is to connect a digital signal to the voltage input. Then, after acquisition, information such as period, frequency and the elapsed times of various logic transitions could be determined. Also, an applications module which records eight digital signals at a programmable rate up to 1 MHz could be built along lines suggested in Section 5.6.

The example problem here is to determine voltage once each millisecond for 1 full second as *accurately* as possible. (It's assumed the voltage doesn't change significantly during several milliseconds.) One approach is to program the FVM's voltage clock for 1000 Hz and 1000 conversions. But greater accuracy is achieved by making 25,000 conversions at a rate of 25 KHz and then successively averaging sets of 25 voltages. Each average covers an elapsed time of 1 ms.

To establish VCK! at 25 KHz and NCV at 25,000, CTR_0 and CTR_1 are loaded with 40 and 25,000, respectively. The following table gives the mode, index, control word and low and high load bytes for all three PIT counters.

counter	mode	index	CW	LOW	HIGH	function
0	3	0,0	&H36	40	0	generate VCK
1	0	0,1	&H70	168	97	end conversions
2	1	0,2	&HB2	1	0	generate SVC*

Starting with the shell FV.BAS, the statements below carry out the following operations: (1) establish arrays including V!(I) for the average voltages; (2) assign values to constants including those in the table above; (3) program the PIT; (4) initiate hardware conversion and storage; (5) wait while M_0 is high; and (6) send stop to the hardware after acquisition.

```
'       Program: V VS T
        DEFINT A-Z
        DIM SELPP(8),PROGP(8)
        DIM SELPC(8,3),LOADPC(8,3),READPC(8,3)
        DIM LOW(7,3),HIGH(7,3),CW(7,3)
        DIM V!(1100)
        COMMON SHARED CP,DP,NORM,STRT,SSTP
        COMMON SHARED SELPP(),PROGP(),SELPC()
        COMMON SHARED LOADPC(),READPC()
        COMMON SHARED MAADV,MALOAD,MREAD,MWRITE,MENDW
        CP = &H300 : DP = &H301
        NORM = &HFF
        STRT = &H46 : SSTP = &HCE
        CW(0,0) = &H36 : LOW(0,0) = 40 : HIGH(0,0) = 0
        CW(0,1) = &H70 : LOW(0,1) = 168
        HIGH(0,1) = 97
        CW(0,2) = &HB2 : LOW(0,2) = 1 : HIGH(0,2) = 0
        SELPP(0) = &HC7
        PROGP(0) = &HC5
        SELPC(0,0) = &H07 : SELPC(0,1) = &H47
        SELPC(0,2) = &H87
        LOADPC(0,0) = &H05 : LOADPC(0,1) = &H45
        LOADPC(0,2) = &H85
        MAADV = &H7F : MALOAD = &HBF
        MREAD = &HFB : MWRITE = &HFD : MENDW = &HC7
MAIN:   CALL SSTOP
        CLS
        FOR J = 0 TO 2
          CALL PROGPIT(0,CW(0,J))
          CALL LOADPIT(0,J,LOW(0,J),HIGH(0,J))
        NEXT J
        PRINT "V vs T for 1.0 Second"
        INPUT "Type Any Key to Start Collection: "; A$
        CALL ZEROADR
        CALL START
WWH:    IF (INP(CP) AND &H01) = 1 THEN GOTO WWH
        CALL SSTOP
```

The next operations are to read the stored binary data from the FVM's memory and determine the average of successive sets of 25 values.

```
        PRINT "Acquisition Over, Computations"
        CALL ZEROADR
        CONV! = 10/256
        FOR I = 1 TO 1000
```

```
      SUMB = 0
      FOR J = 0 TO 24
        CALL READMEM(SVOL)
        CALL ADVADR
        SUMB = SUMB + SVOL
      NEXT J
      V!(I) = SUMB * CONV! / 25
    NEXT I
```

CONV! is the same as in the test/demonstration program. SUMB is the **SUM** of 25 **B**inary voltages. SVOL is a **S**tored **VOL**tage value. And the average **V**oltages are stored in V!(I). The J loop reads and sums the next 25 values. The I loop stores the average voltages. The total number of CALL READMEM(SVOL)'s is 1000 * 25 = 25,000.

The final operation is to display the average voltages in a reasonable format. Since conversions are 1/25,000 seconds apart, the time for each set of 25 is 1 ms which makes the elapsed time to the end of the Ith average I * .0010 second. Fourteen values are on each row. After every 20 rows, display pauses until any key is typed.

```
CLS
K = 0
PRINT "Av. voltage, each value is 1 ms later"
FOR I = 1 TO 1000 STEP 15
  FOR J = 0 TO 14
    PRINT USING "#.##"; V!(I + J);
    PRINT " ";
  NEXT J
  PRINT
  K = K + 1
  IF K > 20 THEN
    PRINT
    INPUT "Type Any Key for the Next Page"; A$
    CLS
    K = 0
  END IF
NEXT I
INPUT "Enter R to Repeat, Else Ends "; A$
IF A$ = "R" OR A$ = "r" THEN GOTO MAIN
END
```

The memory subroutines follow including SSTOP and the modified START.

10.8 Construction

The FVM circuit may be constructed on a wire wrap card identical to the one used for the CDI, ADC and CMS modules. This section lists components, suggests a layout and gives construction details. Section 10.9 outlines testing and troubleshooting.

10.8.1 FVM Components

The fast voltage circuits in Figures 10.2 and 10.13 require the following components.

(1)   One 8253-5 PIT and one 24-pin wire wrap socket.
(2)   Four 6264 memory IC's and four 28-pin wire wrap sockets.
(3)   Four 74LS193 counters and four 16-pin wire wrap sockets.
(4)   One 74LS04 INVERTER, one 74LS08 AND, one 74LS32 OR and one 74LS74 flip-flop.
(5)   One 1.0000 MHz TTL compatible digital oscillator.
(6)   Five 14-pin wire wrap sockets.
(7)   One modular phone jack for EX-S (as described in Section 8.2).
(8)   Telephone cable to connect the jack to experimental apparatus.
(9)   A 40-conductor socket and ribbon cable to connect the CDI and FVM modules (as described in Section 7.7.1).
(10) A 16-conductor socket and ribbon cable to connect the FVM and ADC modules (as described in Section 8.2).
(11) A wire wrap card identical to the CDI's (as described in Section 7.7.1).
(12) A bag of individual wire wrap terminals.
(13) A wire wrap tool and several spools of different color wire.
(14) Six .01 μf disc capacitors.
      Vendors for the components are in Appendix B.

10.8.2 Wire Wrap Card Layout

Figure 10.14 gives a layout for the FVM wire wrap card. If there is a ground plane, it should be on the top or components side. Solder and/or epoxy the fourteen IC and two connector sockets approximately as shown. Epoxy the phone plug. Solder four wire wrap terminals in front of the plug as shown. Solder from left to right the yellow, green, red and black wires to the tops of the terminals. Finally, solder six pairs of terminals up the left side of the card approximately as shown. Solder .01 μf disc capacitors between the terminals. If the card has a ground plane, solder the ground terminals to the plane.

The four 6264's are labeled A, B, C and D as are the four counters. With only one each of the other IC's, they are not labeled. However, 7404, 7408, 7432 and 7474 gate numbers are included in Figures 10.2 and 10.13 and in the wire wrap drawings which follow.

10.8.3 Wire +5, Ground and $d_7 \ldots d_0$

The following sections are a guide for wire wrapping the FVM circuit. The drawings follow the style introduced in Chapter 7 and are useful for both construction and testing.

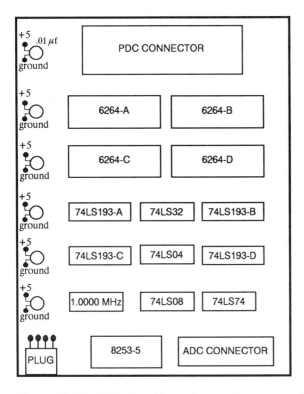

Figure 10.14. FVM circuit layout on a wire wrap card.

Figure 10.15 is the first drawing. The view is from the top or components side. Refer to the actual circuit and make the following connections.

(1) Wire +5 from the top pair of wire wrap terminals to **H** on the CDI connector. Wire ground to both **L**'s on the connector.

(2) Wire +5 from the next lower pair of terminals to the **H**'s on 6264-A and B. Wire ground to the **L**'s on the same chips (avoiding loops).

(3) Wire +5 from the third pair of terminals to the **H**'s on 6264-C and D. Wire ground to the **L**'s.

(4) Wire +5 from the fourth pair of terminals to all **H**'s on 74LS193-A, 74LS32 and 74LS193-B. Wire ground to all **L**'s.

(5) Wire +5 from the fifth pair of terminals to all **H**'s on 74LS193-C, 74LS04 and 74LS193-D. Wire ground to all **L**'s.

(6) Wire +5 from the bottom pair of terminals to all **H**'s on the clock, 74LS08, 74LS74, phone plug, 8253-5 and ADC connector. Wire ground to all **L**'s.

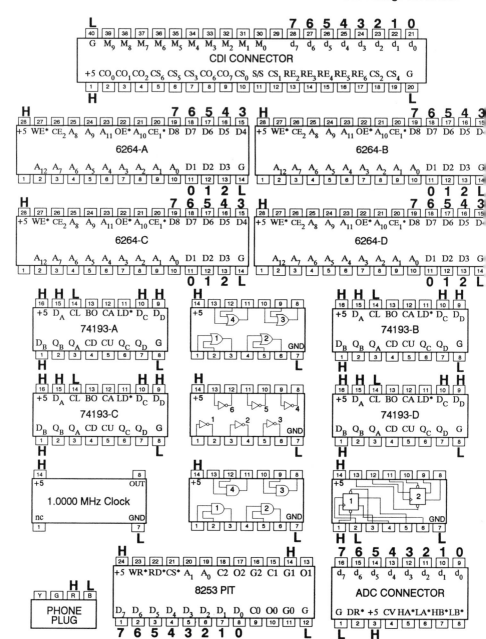

Figure 10.15. FVM wire wrap drawing showing +5, ground and $d_7...d_0$ connections.

(7)   Finally, wire the data lines (designated **0**...**7**) from the CDI connector to all four memory chips, the PIT and the ADC connector. Specifically, connect all pins labeled **0**, etc.

10.8.4 Wire the Memory/address Generator

Figure 10.16 is a guide to wiring the memory/address generator. Proceed as follows.

(1)   The 14 address lines are designated by the corresponding hex digits: **0**...**9, A, B, C, D** and **E**. Wire **0, 1, 2** and **3** from 74LS193-A to all four 6264's.
(2)   Wire **4, 5, 6** and **7** from 74LS193-B to all four 6264's.
(3)   Wire **8, 9, A** and **B** from 74LS193-C to all four 6264's.
(4)   Wire **C** from 74LS193-D to all four 6264's.
(5)   Wire **D** from 74LS193-D to INVERTER-5 and to CE_2 of 6264-B and D.
(6)   Wire **E** from 74LS193-D to INVERTER-6 and to $CE_1{}^*$ of 6264-A and B.
(7)   Wire INVERTER-5's output **M** to CE_2 of 6264-A and C.
(8)   Wire INVERTER-6's output **N** to $CE_1{}^*$ of 6264-C and D.
(9)   Wire the CDI connector's CO_2 (**P**) to OE* on all four 6264's.
(10) Wire AND-4's output **Q** to WE* on all four 6264's.
(11) Wire the CDI connector's CO_6 (**R**) to LD* of all four counters and to the PIT's A_0.
(12) Wire OR-1's output **S** to 74LS193-A's CU input. Wire 74LS193-A's CA output **T** to 74LS193-B's CU. Wire 74LS193-B's CA output **U** to 74LS193-C's CU. And wire 74LS193-C's CA output **V** to 74LS193-D's CU.
      This completes the memory/address generator.

10.8.5 Wire the PIT and ADC Connector

Figure 10.17 is a guide for wiring the remainder of the circuit. Proceed as follows.

(1)   Wire the CDI connector's $CS_0{}^*$ (**A**) to the PIT's CS*.
(2)   Wire the CDI's $CO_1{}^*$ (**B**) to the PIT's WR*.
(3)   Wire the CDI's $CO_2{}^*$ (**C**) to the PIT's RD* (although the line is never used).
(4)   Wire the CDI's $CO_7{}^*$ (**D**) to the PIT's A_1. (A_0 was wired in the previous section.)
(5)   Wire the CDI's S/S* (**E**) to AND-1.
(6)   Wire INVERTER-1's output **G** to AND-2, AND-3 and the CDI's M_0.
(7)   Wire the clock's output **I** to the PIT's CK_0 and CK_2.
(8)   Wire phone plug output **J** to AND-1. This is EX-S.

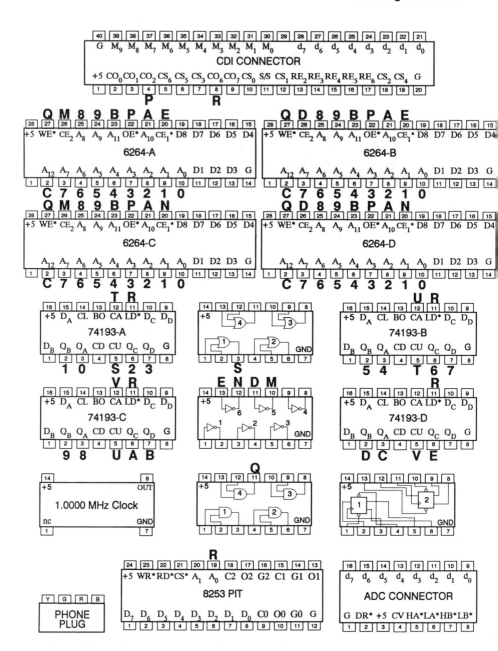

Figure 10.16. FVM wire wrap drawing showing memory/address generator connections.

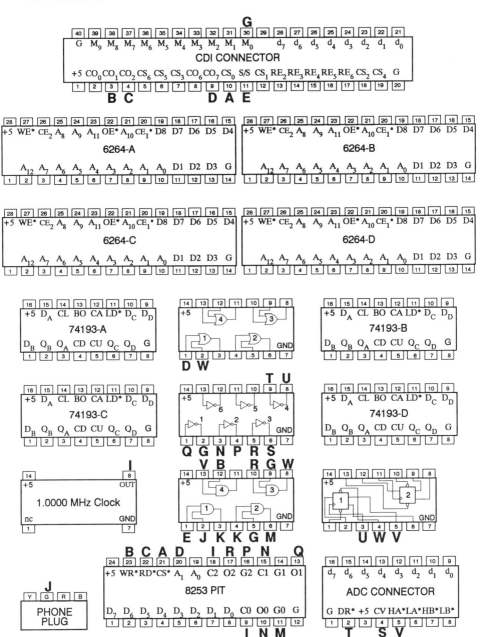

Figure 10.17. FVM wire wrap drawing for the remainder of the circuit.

(9)   Wire AND-1's output **K** to AND-2.

(10) Wire AND-2's output **M** to the PIT's G_0.

(11) Wire the PIT's OUT_0 (**N**) to CK_1 and INVERTER-2.

(12) Wire INVERTER-2's output **P** to G_2.

(13) Wire the PIT's OUT_1 (**Q**) to INVERTER-1's input.

(14) Wire the PIT's OUT_2 (**R**) to AND-3 and INVERTER-3.

(15) Wire INVERTER-3's output **S** to the ADC connector's CV.

(16) Wire the ADC connector's DR* (**T**) to INVERTER-4.

(17) Wire INVERTER-4's output **U** the FF-1's CK input.

(18) Wire FF-1's output **V** to the ADC connector's HA* and to AND-4.

(19) Wire AND-3's output **W** to OR-1 and FF-1's PRESET*.

This completes the fast voltage measurer circuit. Visually inspect all work. Install the IC's.

10.9 Testing and Troubleshooting

In order to test the FVM circuit and software, assemble the system as follows:

(1)   Connect the parallel ports to the CDI module with appropriate cable(s).

(2)   Connect the CDI to the FVM module with a 40-conductor ribbon cable.

(3)   Connect the FVM to the ADC module with a 16-conductor ribbon cable.

(4)   Connect the power supply to the ADC module with a phone cable.

Before beginning, it's essential that the parallel ports, CDI and ADC modules work. If necessary, repeat the test procedures in earlier chapters. Troubleshooting tools are a logic probe, a multimeter, a variable frequency digital clock, an oscilloscope and a set of IC test clips (14, 16, 24 and 28 pin).

10.9.1 Continuity Tests

With the ADC and CDI modules temporarily disconnected, test FVM wiring by confirming that the resistance between pins which should be connected is zero. Put multimeter probes directly on IC pins and into connector sockets. Use Figures 10.15-10.17 as a guide and follow the order used in construction. The most common error is omitting a connection.

10.9.2 Test +5 and Ground

Connect the ADC module and turn the power on. Use the multimeter to measure +5 volts. If the value isn't at least 4, disconnect the FVM and measure +5 in the ADC module. If correct there, remove the IC's from the FVM, reconnect the ADC and measure again. If the voltage is still incorrect, look for a solder bridge or bad power supply noise capacitor. If correct after removing the IC's, systematically reinstall the chips to identify the bad one.

Check all +5 and ground connections in Figure 10.15 with a logic probe.

10.9.3 Test PIT Programming and Loading

The first operational tests are PIT programming and loading. Starting with the CDI software shell in Chapter 7, set up the modified shell FV.BAS described in Section 10.6.1. Print, carefully proof and correct as necessary.

Remove CALL STOP and enter the following main program.

```
MAIN:  CALL PROGPIT(0,&H70)
       GOTO MAIN
```

Start execution and then use an oscilloscope to systematically verify that all PIT controls follow the timing pattern in Figure 6.10. Connect scope probes via an IC test clip directly to 8253 pins. With the scope triggered on WR*, check each data bit for 70 hex. The cause of an incorrect control or data signal could be either a wiring error or an incorrect software constant. After PROGPIT checks out, use Figure 6.11 to test LOADPIT with the following main program.

```
MAIN:  CALL LOADPIT(0,0,0,0)
       GOTO MAIN
```

Recall that an infinite QuickBASIC loop is exited by typing CTRL-BREAK.

In both program and load operations, it's possible that WR* is not active the required 400 ns (as discussed in Section 6.4.2). If necessary, extend the signal by adding OUT statements in the subroutines.

When the program and load routines check out, enter this main program.

```
MAIN:  CALL SSTOP
       CALL PROGPIT(0,&H36)
       CALL LOADPIT(0,0,16,39)
       CALL PROGPIT(0,&H70)
       CALL LOADPIT(0,1,256,256)
       CALL PROGPIT(O,&HB2)
       CALL LOADPIT(0,2,100,0)
       CALL START
       END
```

The statements program PIT counters 0, 1 and 2 for modes 3, 0 and 1, respectively. CTR_0 is loaded with 10,000 which makes VCK 100 Hz. CTR_1 is loaded with 64K to produce the maximum number of conversions and thereby give the most time to evaluate the circuit before FV-E* goes low. CTR_2 is loaded with 100 which makes the convert pulse 100 μsec long and easier to see on an oscilloscope.

Run the program. Check G_0 with the logic probe. It should be high. If not, check S-S/S* for high. If the parallel ports and CDI module work and S-S/S* is low, a software constant is wrong. If S-S/S* is correct and G_0 wrong, check FV-E* which should also be high. If it is, the problem must be AND gates 1 and 2.

On the other hand, if FV-E* is incorrectly low, check OUT_1. If high as expected, the problem is INVERTER-1. If OUT_1 is wrong, the PIT is incorrectly programmed or defective. Replace the 8253 and repeat the control and data tests above.

When G_0, OUT_1 and FV-E* check out, use the oscilloscope to verify that OUT_0's frequency is 100 Hz. If it isn't, check the 1.0000 MHz digital clock as it reaches CK_0.

Next, verify with the scope that OUT_2 is a 100 μsec low pulse occurring once each OUT_0 cycle. If it isn't, check CK_2 and VCK as they reach CTR_2.

If the system works as designed, after 64K VCK cycles, OUT_1 should go high and FV-E* low. Since VCK's period is .010 second, this occurs after 640 seconds. Watch for FV-E* low and rerun the program if necessary.

10.9.4 Test the ADC and Control Flip-flop

Run the main program above. Connect oscilloscope channel 1 via an IC test clip to CV (at INVERTER-3's output) and connect DR* to channel 2 (at INVERTER-4's input). DR* should go high after CV and return low 20 μsec later. (Recall that CV is programmed for 100 μsec, not the usual 1 μsec). If DR* is wrong, check the ADC cable and then test the ADC module using procedures at the end of Chapter 8.

Still triggering on CV connected to channel 1, observe SVC'* on channel 2. It should be the inverse of CV. Finally, observe the flip-flop's output SIM*. It should go low when data is ready 20 μsec after CV and return high at the beginning of the next conversion 10,000 μsec later. (Put another way, SIM* consists of a brief high pulse at the beginning of each VCK cycle.) If SIM* is incorrect, check that the flip-flop's DATA and CLEAR* inputs are low and high, respectively.

The goal of the these tests is to isolate something which is definitely wrong. Once a problem is clear, it's usually straightforward to find the cause. As always, patience and persistence are required.

10.9.5 Test the Memory/Address Generator

For software to interact with the memory/address generator, SVC'* and SIM* must be low and high, respectively. However, as discussed at the end of Section 10.5, unless CTR_1 has decremented to zero and made FV-E* low, SVC'* may be high (which would prevent software CU from reaching the low counter). A way to avoid the situation is to change the current test program so CTR_1 is loaded with 100 rather than 64K. With VCK at 100 Hz, CTR_1 decrements to zero in 1 second at which point FV-E* goes low. This makes SVC'* and SIM* definitely low and high, respectively. So change the main program to:

```
MAIN:  CALL SSTOP
       CALL PROGPIT(0,&H36)
       CALL LOADPIT(0,0,16,39)
       CALL PROGPIT(0,&H70)
       CALL LOADPIT(0,1,100,0)
       CALL PROGPIT(O,&HB2)
       CALL LOADPIT(0,2,100,0)
       CALL START
       END
```

Check Zero Address

Add the following statements after CALL START:

```
       INPUT "Check SVC'* and SIM*, Then Any Key ";A$
TLOOP: CALL ZEROADR
       INPUT "Check the Addresses ";A$
       END
```

Run the program. When the input occurs, check with the logic probe that SVC'* is low and SIM* is high. There should be no problem if the previous checks were successful. After any key is typed and the second input occurs, check all address bits with the logic probe. If any differ from the expected zero, check each counter's LD* (high), data inputs (all high) and CLear (low).

If no problem is found but one or more address bits are still wrong, put CALL ZEROADR in an infinite loop and check LD* with an oscilloscope at all four counters' pin 11. Then check both LD* and CU on counter-A.

Check Advance Address

Once ZEROADR works, replace the three statements above with:

```
TLOOP: CALL ADVADR
       GOTO TLOOP
       END
```

Run the program and observe successively higher address bits with the logic probe on 6264-A's pins. Eventually, the transitions are slow enough to see. After that, each higher bit should change less rapidly. Use the oscilloscope to observe the lower address bits using the 28-pin test clip on 6264-A. If ZEROADR previously checked out, a problem is unlikely here.

Check Write to Memory and Read from Memory

Replace the two statements above with the following (still immediately after CALL START):

```
       INPUT "Check SVC'*, Then Any Key ";A$
       CALL ZEROADR
       FOR I = 0 TO 25
```

```
      CALL WRITEMEM(I)
      CALL ADVADR
NEXT I
CALL ENDMEMWR
CALL ZEROADR
FOR I = 0 TO 25
      CALL READMEM(DVALUE)
      PRINT DVALUE;"    ";
      CALL ADVADR
NEXT I
END
```

Run the program and see if values from 0 to 25 are displayed. If not, separately put WRITEMEM and READMEM in infinite loops and use an oscilloscope to verify the timing patterns in Figures 5.2 and 5.3. Again, likely problems are unmade connections and software constant errors.

When the memory checks out, restore the original test/demonstration program. (It is a good idea to save the various troubleshooting programs for later use.)

10.9.6 Run the Test/demonstration Program

Connect a known, constant voltage to ADC573-A, perhaps from a battery. Run the test/demonstration program at some rate for some number of conversions and see if correct values are found. Another interesting measurement is to connect a 1000 Hz digital clock to voltage input-A. Then, run the program at 25,000 Hz for 200 conversions. If the system is working, the voltages should be $+5$ or so and zero or so and there should be 25 values per cycle.

10.10 Conclusions

Chapters 9 and 10 describe two measurement systems which work through the parallel ports and which use the ADC and CDI modules. The systems illustrate techniques for digital design, software development and construction and testing. The techniques are applicable to a variety of additional measurement needs.

Exercises

1.   What happens in the FVM circuit if either EX-S or S-S/S* go low during hardware measurement?

2. Design an FVM circuit in which EX-S properly stops a conversion sequence by going low. After a stop, software reads CTR_1 to determine the number of conversions up to that point. Sketch the circuit. Change the test/demonstration program to accommodate the new capability.

3. Devise a way to use A_{13}, A_{14} and A_{15} from the high address counter to operate the eight 6264's needed for a 64K memory. A possible approach uses a 74138 and additional AND and OR logic. Sketch the circuit. Indicate which 6264 responds to which 8K memory block. Give any software changes necessary to operate the expanded memory.

4. The problems in going beyond the eight 6264's in Exercise 3 include CTR_1's 64K capacity and the need for another address generating counter. Outline solutions to both problems.

5. In Figure 10.11, the first conversion is delayed until VCK goes high one-half of an OUT_0 cycle after start. Design circuitry which initiates the first conversion at the beginning of the first OUT_0 cycle so the maximum delay after start is 1 and 1/2 cycles of the 1.0000 MHz clock. Give a complete timing diagram and sketch the circuit.

6. Write a program for FVM hardware which repeatedly displays the average of 100 voltage readings. Start with FV.BAS in Section 10.6.1.

7. Suppose a digital clock is applied to voltage input A. The maximum and minimum frequencies are 10,000 and 100 Hz. Write a program which repeatedly determines and displays the clock's frequency with 1% accuracy. There are two approaches. In one, the number of conversions in 100 cycles is counted and frequency determined from the elapsed time. In the other, the number of conversions in one cycle is counted and the period determined. Software can switch between approaches and, of course, can reprogram the conversion rate depending on circumstances. Start with FV.BAS in Section 10.6.1 and give the program.

8. Design an FVM type circuit which determines voltage at 10-bit accuracy up to 16K times. Use an additional PIT and/or a second address generator. Sketch the circuit. Start with FV.BAS in Section 10.6.1 and write a test/demonstration program. Include any revised subroutines.

9. Design hardware for a new FVM which simultaneously determines voltage at 8-bit accuracy from both AD573's with up to 16K values from each. Use an additional PIT and/or address generator as necessary. Sketch the circuit. Start with FV.BAS in Section 10.6.1 and write a test/demonstration program. Include any revised subroutines.

10. Design hardware for an applications module which records up to eight digital signals with the rate and number programmable up to 500 KHz and 32K, respectively. Start with the fast recorder in Section 5.6 and add PIT control. Give a detailed timing diagram. Sketch the circuit. Start with FV.BAS in Section 10.6.1 and write a test/demonstration program.

11. In another memory application, software downloads a sequence of values. Afterward, hardware sequentically makes the values available to outside circuitry at a programmable rate. Design an applications module which accomplishes this for a maximum rate of 500 KHz and for up to 32K values. Include a trigger to initiate the read memory sequence. Give a complete timing diagram. Sketch the circuit. Start with FV.BAS in Section 10.6.1 and write a test/demonstration program.

12. What are the problems in making the "playback" sequence in Exercise 11 repeat a programmed number of times?

13. What are the problems in integrating the digital recorder in Exercise 10 with the playback system in Exercise 11?

14. Consider the Analog Devices AD558 8-bit digital-to-analog converter (DAC). Use the *Data-Acquisition Handbook* (reference in Appendix A) to design an applications module in which software downloads up to 32K values to the memory after which hardware makes the stored values available to the DAC at a programmable rate. Include a hardware trigger for the playback sequence. Give a complete timing diagram. Sketch the circuit. Start with FV.BAS in Section 10.6.1 and write a test/ demonstration program.

15. What are the problems in integrating the voltage playback system in Exercise 14 with the FVM described in this chapter?

16. Construct and test the FVM circuit. Use guidelines in Sections 10.8 and 10.9 as desired.

17. Enter the test/demonstration program in Section 10.6. Evaluate the fastest possible hardware conversion rate using suggestions in Section 10.6.3.

Appendix A References

General Electronics

Horowitz, P. & Hill, W., *The Art of Electronics*, Cambridge Univ. Press, New York, 1980.

Higgins, R., *Electronics with Digital and Analog Integrated Circuits*, Prentice Hall, Englewood Cliffs, 1983.

Millman, J., *Microelectronics*, McGraw-Hill, New York, 1979.

Digital Electronics

Greenfield, J., *Practical Digital Design Using IC's*, Wiley, New York, 1977.

Wagner, D., *Digital Electronics*, Harcourt Brace Jovanovich, San Diego, 1988.

Analog Electronics

Faulkenberry, L., *An Introduction to Operational Amplifiers*, Wiley, New York, 1977.

Computer Interfacing and Programming

Titus, J., Larsen, D. & Titus, C., *Apple Interfacing*, Blacksburg Group, Blacksburg, 1981.

Robbins, J., *Understanding Microsoft QuickBASIC*, Sams, Indianapolis, 1988.

Sensors, Transducers and Computers in the Laboratory

Snider, J. & Priest, J., *Electronics for Physics Experiments Using the Apple II Computer*, Addison-Wesley, Reading, 1989.

Kuckes, A. & Thompson, B., *Apple II in the Laboratory*, Cambridge Univ. Press, New York, 1987.

Gates, S. & Becker, J., *Laboratory Automation Using the IBM PC*, Prentice Hall, Englewood Cliffs, 1989.

Ratzlaff, K., *Introduction to Computer-Assisted Experimentation*, Wiley, New York, 1987.

Warring, R. & Gibilisco, S., *Fundamentals of Transducers,* Tab, Blue Ridge Summit, 1985.

Tien Lang, T., *Electronics of Measuring Systems,* Wiley, New York, 1987.

Horn, D., *How to Use Special-Purpose ICs*, Tab, Blue Ridge Summit, 1986.

General Reference Manuals

The TTL Data Book for Design Engineers, Texas Instruments Incorporated, P. O. Box 5012, Dallas, TX, 75222.

Linear Databook, Vol. 1, 2, 3, National Semiconductor Corporation, 2900 Semiconductor Drive, Santa Clara, CA, 95052.

Specific Reference Manuals

Apple IIe Reference Manual, Apple Computer.

Technical Reference for the Personal Computer AT, IBM Corporation.

Data-Acquisition Databook, Vol. 1, Analog Devices, P.O. Box 280, Norwood, MA, 02062.

CMOS Memories, DRAMS & SRAMS, Vitelic Corporation, 3910 North First Street, San Jose, CA, 95134.

IC Memories Data Book, Hitachi America, Ltd., 1800 Bering Drive, San Jose, CA, 95112.

Microsystems Components Handbook, Microprocessors and Peripherals (Vol. I), Intel Corporation, 3065 Bowers Avenue, Santa Clara, CA, 95051.

Appendix B   Vendors

General Electronics Suppliers

Jameco Electronics
1355 Shoreway Road
Belmont, CA 94002

Allied Electronics
401 E. 8th Street
Fort Worth, TX 76102

Digi-Key Corporation
701 Brooks Ave. South
Thief River Falls, MN 56701

Newark Electronics
500 North Pulaski Road
Chicago, IL 60624

MCM Electronics
650 Congress Park Dr.
Centerville, OH 45459

Mouser Electronics
P.O. Box 699
Mansfield, TX 76063

Specific Device Suppliers

AD573 ADC's
Analog Devices
Two Technology Way
Norwood, MA 02062

Right-angle Connector for IBM Parallel Ports
Norwesco, Inc.
Gastonia, NC 28054

Breadboard Systems
E&L Instruments, Inc.
61 First Street
Derby, CT 06418

Parallel I/O Ports
See pages 72-73

Appendix C   Pin Assignments

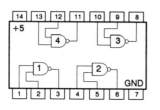

7400 NAND

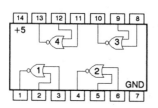

7402 NOR

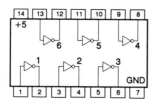

7404 INVERTER

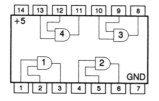

7408 AND

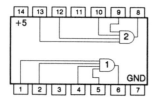

7421 4-Input AND

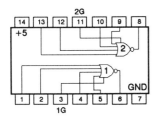

7425 4-Input NOR
(wire gates high)

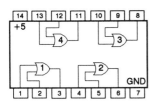

7432 OR

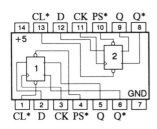

7474 Flip-flop

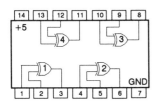

7486 EXCLUSIVE-OR

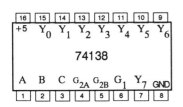

74138 Decoder

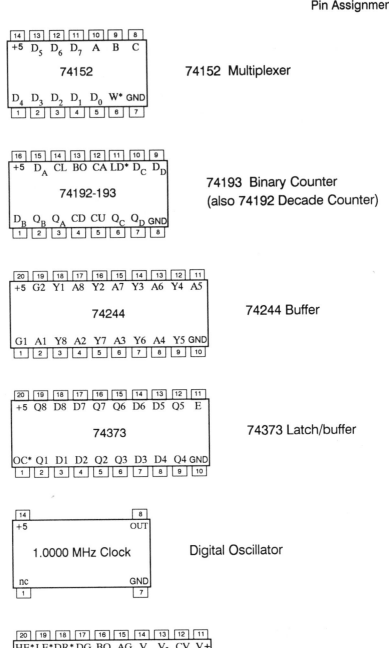

74152 Multiplexer

74193 Binary Counter
(also 74192 Decade Counter)

74244 Buffer

74373 Latch/buffer

Digital Oscillator

AD573 Analog-to-digital
Converter

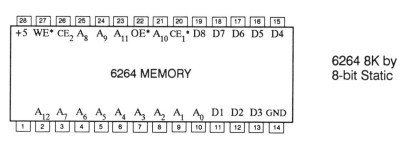

28	27	26	25	24	23	22	21	20	19	18	17	16	15
+5	WE*	CE_2	A_8	A_9	A_{11}	OE*	A_{10}	CE_1*	D8	D7	D6	D5	D4

6264 MEMORY

A_{12}	A_7	A_6	A_5	A_4	A_3	A_2	A_1	A_0	D1	D2	D3	GND
1	2	3	4	5	6	7	8	9	10	11	12	13	14

6264 8K by
8-bit Static

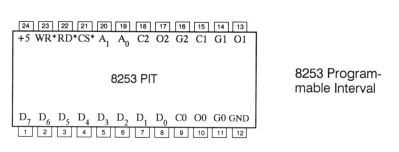

24	23	22	21	20	19	18	17	16	15	14	13
+5	WR*	RD*	CS*	A_1	A_0	C2	O2	G2	C1	G1	O1

8253 PIT

D_7	D_6	D_5	D_4	D_3	D_2	D_1	D_0	C0	O0	G0	GND
1	2	3	4	5	6	7	8	9	10	11	12

8253 Program-
mable Interval

34	33	32	31	30	29	28	27	26	25	24	23	22	21	20	19	18
CI_7	CI_5	CI_3	CI_1	CO_7	CO_5	CO_3	CO_1	G	DI_7	DI_5	DI_3	DI_1	DO_7	DO_5	DO_3	DO_1

PARALLEL PORTS CONNECTOR

CI_6	CI_4	CI_2	CI_0	CO_6	CO_4	CO_2	CO_0	G	DI_6	DI_4	DI_2	DI_0	DO_6	DO_4	DO_2	DO_0
1	2	3	4	5	6	7	8	9	10	11	12	13	14	15	16	17

Connector End of Parallel Ports Cable (as recommended in Appendices D and E)

40	39	38	37	36	35	34	33	32	31	30	29	28	27	26	25	24	23	22	21
G	M_9	M_8	M_7	M_6	M_5	M_4	M_3	M_2	M_1	M_0		d_7	d_6	d_5	d_4	d_3	d_2	d_1	d_0

CDI CONNECTOR

+5	CO_0	CO_1	CO_2	CS_6	CS_5	CS_3	CO_6	CO_7	CS_0	S/S	CS_1	RE_2	RE_3	RE_4	RE_5	RE_6	CS_2	CS_4	G
1	2	3	4	5	6	7	8	9	10	11	12	13	14	15	16	17	18	19	20

Connector Between the CDI and Applications Modules (as recommended in Chapter 7)

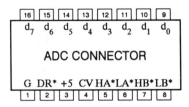

Connector Between the Applications and ADC Modules (as recommended in Chapter 8)

Appendix D   IBM Parallel Ports

Section 3.2 explains the design of a pair of 8-bit input/output ports for IBM PC/XT/AT and compatible computers. Figure 3.11 is the complete circuit. This appendix is a guide to building the ports. Section D.1 lists components and presents a layout. Section D.2 suggests an interfacing cable and connectors. Section D.3 gives wiring instructions. And Section D.4 outlines testing and troubleshooting. The objective is a complete presentation of one construction approach. Details such as connectors and cables are somewhat arbitrary and other schemes may be substituted.

D.1 Components and Layout

The circuit in Figure 3.11 requires the following components.
(1)   A Vector Model 4613-1 or 4617-1 wire wrap board for the IBM PC/XT or PC-AT expansion slot.
(2)   Four 74LS373 latch/buffers.
(3)   Two 7425 4-input NOR IC's.
(4)   Two 74LS04 INVERTER chips.
(5)   One 74LS138 decoder.
(6)   Nine IC sockets: four 20-pen, one 16-pin, and four 14-pin.
(7)   A bag of Vector T44 wire wrap terminals.
(8)   Six .01 μf disc capacitors.
(9)   A wire wrap tool and an assortment of wire.
Vendors for these components are in Appendix B.

Figure D.1 suggests a layout for the circuit on a PC-AT expansion board. The view is from the components side with the rear of the computer on the right. (The regular PC board differs primarily in the absence of the second plug on the bottom left.)

Along the bottom, just above the right plug, are holes for wire wrap terminals. When inserted, the terminals allow expansion slot signals to be connected to the ports circuit. The lower row starts with A1 on the right and ends on the left with A31. The upper row starts on the right with B1 (which is ground) and ends with B31.

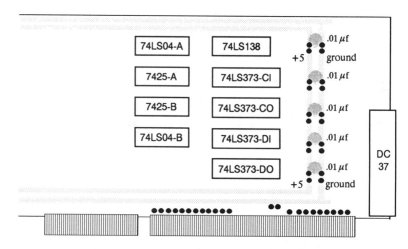

Figure D.1. Layout for the I/O ports circuit on a Vector PC-AT expansion board. The view is from the components side and the rear of the computer is on the right. The board extends further to the left than shown.

The following steps prepare the board for wiring.

(1)   Solder wire wrap terminals to the following expansion slot holes: A2 - A9, A11, A20 - A31, B13 and B14. Dots in Figure D.1 show the approximate locations. A terminal is pushed through from the components to the solder side.

(2)   Epoxy the nine wire wrap sockets approximately as shown in the layout.

(3)   Finally, place pairs of wire wrap terminals along the vertical +5 and ground strips to the right of each row of sockets. In holes just above the terminals, solder .01 μf capacitors between +5 and ground (for power supply noise reduction). The +5 and ground strips are connected to the computer's power supply and ground via the expansion board's plug.

D.2 Connectors and Cable

It's necessary to construct a cable to run from the ports (in the computer) to various applications on breadboards and wire wrap cards. Recall from Chapter 3 that the two 8-bit I/O ports present 32 connections. Since ground is also required, a 34-conductor ribbon cable is selected. (Actually, for reasons explained below, 37-conductor cable is used.)

A common practice is to use multiple ground lines, sometimes every other wire in circuits with high frequency signals. Extra grounds have proven unnecessary here even for 20-foot cables.

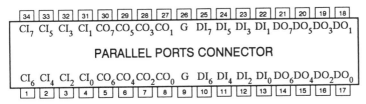

Figure D.2. Pin assignments at the applications end of the parallel ports cable. CI stands for Control-In, CO for Control-Out, DI for Data-In and DO for Data-Out.

A 34-conductor edge socket is chosen for the applications end of the cable. The breadboard or wire wrap card receives the socket with a 34-conductor header. Pin assignments for the receiving header, henceforth called the parallel ports connector, are in Figure D.2. (When used on a breadboard, it's necessary to slightly bend the header's pins.)

Vector expansion slot boards contain pads for right-angle D-subminiature connectors so that when a board is inserted in the computer the connector is easily accessible on the back (the same as commercial boards). Vector supplies a bracket to support the connector and attach the board to the computer's rear. A 37-pin female connector is selected for the board and a matching male connector for the ribbon cable.

The reason for 37-pin connectors is simply because that's the nearest available size greater than 34.

The specific connector and cable components are:
(1) One DC37 female, right-angle, D-subminiature connector for the expansion board (such as Tex-Techs Model P37S-03, .590" series).
(2) One 37-pin, male, D-subminiature, flat cable connector for the computer end of the ribbon cable (such as 3M series 8237).
(3) Some length of 37-conductor ribbon cable.
(4) One 34-conductor, .100" by .100", flat cable edge socket for the other end of the ribbon cable (such as 3M series 3414).
(5) Several .100" by .100", 34-pin, wire wrap type headers for applications circuits (such as 3M series 3594).

Vendors are listed in Appendix B.

Mount the connectors as follows:
(1) Pin assignments for the DC37 right-angle connector are in Figure D.3 (as viewed from the components side). Note that pins 18, 19 and 20 are not used. Also, the pin numbers don't always match those of the parallel ports connector in Figure D.2. Temporarily place the DC37 on the board and note where wire wrap terminals must be placed. Remove

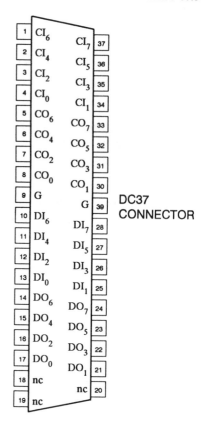

Figure D.3. Pin assignments for the DC37 right angle connector on the expansion board. Pins 18, 19 and 20 are not connected.

the connector and push through and solder 34 terminals. On the AT board, the terminal tops on the components side must be cut so the connector properly fits. Replace and solder the connector. There is no choice in its orientation.

(2) Adjust the cutout in the bracket supplied by Vector to fit the DC37. Mount the bracket.

(3) Mount the male D-connector on one end of the ribbon cable. Connectors designed for flat cable are mounted with a vise. Just observe what has to happen and use common sense.

(4) There is a miss-match between the capacity of the ribbon cable and the 34-conductor edge socket. Strip an inch or so of the three outer wires on the side of the cable corresponding to DC37 pins 18, 19 and 20. Then mount the edge socket.

A good idea is to run the cable between the header on a breadboard and the DC37 and employ a multimeter to test continuity. Use the pin assignments in Figures D.2 and D.3.

D.3 Wiring

The next step is to connect the IC sockets and terminals. Three "wire wrap" drawings, the first in Figure D.4, facilitate the work. They show the DC37 connector, the 23 wire wrap terminals adjacent to the plug, and the 9 IC sockets, all from the components side. The arrangement is similar to the layout in Figure D.1 except the +5 and ground terminals are omitted. The 74LS373 IC's are distinguished according to function. The 7425's are labeled A and B as are the 74LS04's. The 74LS138 is unique. Gates in each 7425 are numbered 1 and 2 and gates in each INVERTER are numbered 1 to 6. In the actual circuit in Figure 3.11, all NOR's and INVERTERS are labeled by IC and gate number. So, NOR-A2 is 7425-A's gate 2.

In each wire wrap drawing, bold characters are placed above and below pins and terminals. Points to be connected are labeled by the same character.

Another approach is to use drawings similar to Figure D.4 but "up-side-down-and-backward." This is because the wire wrap card has to be turned over to make connections. The "right-side-up" method here encourages thinking about each step in terms of what's being wired. And it facilitates testing and troubleshooting as described in Section D.4.

D.3.1 Wire +5 and Ground

Use Figure D.4 to wire +5 to each **H** labeled pin and ground to each **L** pin. Start on the right with the separate +5 and ground wire wrap terminals (not shown in the drawing). Wire across a row of IC's. It's important not to make ground "loops." Included in the **H**'s and **L**'s are IC pins which must be wired high or low as indicated in Figure 3.11. Also, don't forget the **L**'s on the expansion board connector. It's customary to use red wire for +5 and black for ground.

D.3.2 Wire $d_7...d_0$

Connect the eight data lines from the wire wrap terminals adjacent to the plug to all four 74LS373 IC's. Specifically, wire all pins labeled **0** in Figure D.4. Wire all **1**'s, all **2**'s, and so on up to all **7**'s. Note that $d_7...d_0$ is connected to the Q's of the IN 74LS373's and to the D's of the OUT 74LS373's.

D.3.3 Wire Control Circuitry

Refer to Figures 3.11 and D.5 and make the following connections.
(1) Wire address line **0** from its wire wrap terminal to 74LS138 select C.

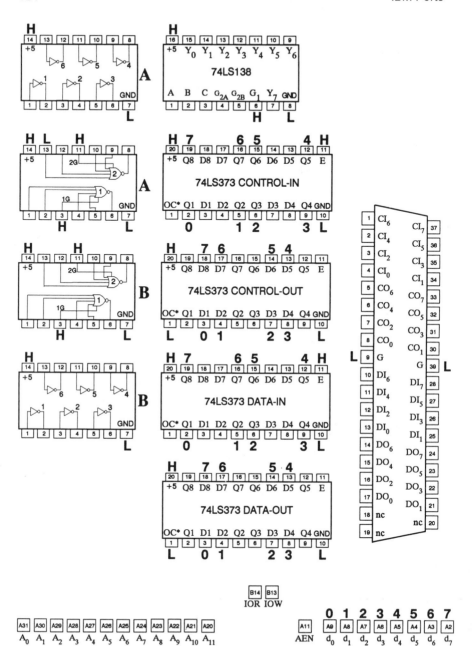

Figure D.4. Wire wrap drawing for the PC ports circuit showing +5 and ground connections as well as $d_7...d_0$.

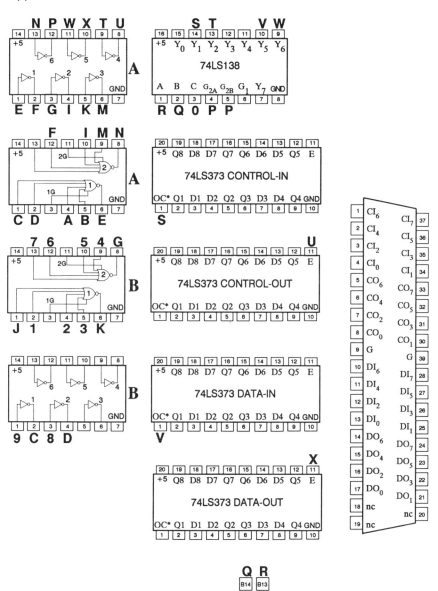

Figure D.5. Wire wrap drawing for port number decoding and control circuitry.

(2) Wire address lines **1**, **2** and **3** from their respective wire wrap terminals to NOR-B1.

(3) Wire address lines **4**, **5**, **6** and **7** to NOR-B2.

(4) Wire address lines **8** and **9** to INVERTER's B2 and B1, respectively.

(5) Wire address A_{10}, labeled **A**, to NOR-A1. Wire A_{11} (**B**) also to NOR-A1.

(6) Wire INVERTER-B1's output **C** to NOR-A1. Wire INVERTER-B2's output **D** also to NOR-A1. Wire NOR-A1's output **E** to INVERTER-A1. Wire INVERTER-A1's output **F** to NOR-A2.

(7) Wire NOR-B2's output **G** to INVERTER-A2. Wire INVERTER-A2's output **I** to NOR-A2.

(8) Connect wire wrap terminal AEN (**J**) to NOR-B1. Wire NOR-B1's output **K** to INVERTER-A3. Wire INVERTER-A3's output **M** to NOR-A2. Wire NOR-A2's output **N** to INVERTER-A6. And, wire INVERTER-A6's output **P** to 74LS138 G_{2A}* and G_{2B}*.

(9) Connect wire wrap terminal IOR (**Q**) to 74LS138 select B. Wire terminal IOW (**R**) to 74LS138 select A.

(10) Wire 74LS138 output Y_1 (**S**) to the control-in 73LS373's OC*.

(11) Wire 74LS138 output Y_2 (**T**) to INVERTER-B4. Wire INVERTER-B4's output **U** to the control-out 74LS373's enable.

(12) Wire 74LS138 output Y_5 (**V**) to the data-in 74LS373's OC*.

(13) Wire 74LS138 output Y_6 (**W**) to INVERTER-A5. Wire INVERTER-A5's output **X** to the data-out 74LS373's enable.

D.3.4 Wire the Expansion Board Connector

Finally, use Figure D.6 to wire the 32 lines from the right-angle DC37 connector to the respective 74LS373's. Specifically:

(1) Wire socket pins labeled **0** to **7** to the control-in 74LS373's D inputs.

(2) Wire socket pins labeled **A** to **H** to the control-out 74LS373's Q outputs.

(3) Wire socket pins labeled **I** to **P** to the data-in 74LS373's D inputs.

(4) Wire socket pins labeled **Q** to **X** to the data-out 74LS373's Q outputs. This completes construction. Visually inspect the work. Install the IC's.

D.4 Testing and Troubleshooting

The following tools are needed to test and troubleshoot the circuit: a logic probe, IC test clips (14-, 16- and 20-pin), a multimeter (which measurers voltage and resistance), an oscilloscope and an extender card for the PC or PC-AT.

As discussed in Chapter 3, the control port number is 300 hex (768 decimal) and the data port is 301 hex (769). It's essential that no other boards

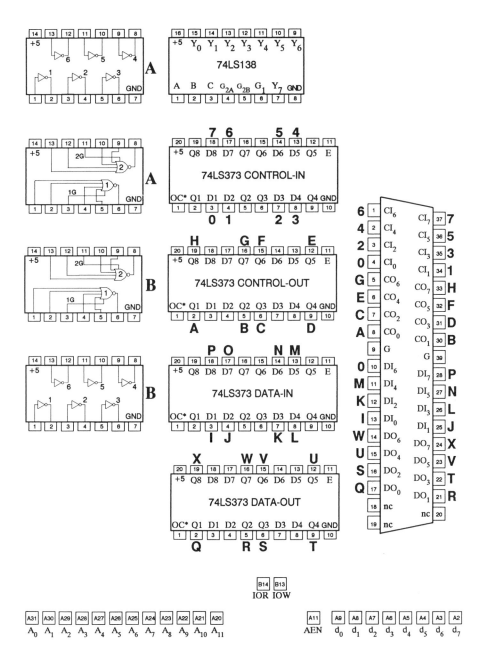

Figure D.6. Wire wrap drawing for connecting the DC37 to the 74LS373's.

have these numbers. Review specifications for all expansion slot devices. Video display, parallel and serial printer, and disk controller circuits should not be a problem. However, unusual boards such as another data acquisition system might be. If an incompatibility is found, the port number decoding circuitry in Section 3.2.2 can be changed.

D.4.1 Continuity Tests

Before installing the circuit, test all wiring by verifying with the multimeter that the resistance between points which should be connected is zero. Use Figures 3.11 and D.4-D.6 and follow the order used in construction. Put multimeter probes directly on IC pins to detect the possibility of a defective socket or bent pin.

If not carried out earlier, run the ribbon cable from the I/O ports to a breadboard. Then, use Figures D.2 and D.3 to test the continuity of all connections.

D.4.2 Test +5 and Ground

The next step is to check +5 and ground. Before doing so, read the computer's manual on installing expansion slot cards. Always follow instructions, one of which is to have the power off when inserting or removing a card.

Open the computer and insert the extender card. Next, install the ports in the extender. Turn the power on. If any chip rapidly heats up, turn the power off. If the computer doesn't boot properly, turn the power off. In either case, remove the ports circuit and repeat the continuity tests. Look for bent wire wrap pins or a solder bridge. If no problem is found, remove the IC's and cable, reinsert the board and turn the power on. The computer should boot. Next, reinstall the IC's one-by-one rebooting each time. When the computer doesn't boot or the just installed IC rapidly heats up, the chip is either defective or incorrectly wired.

Once the computer boots with the ports installed and no problems are apparent, use the multimeter to measure +5 at some point on the expansion board. If it isn't at least 4 volts, turn the computer off and remove the board. Turn the power back on and measure +5 at pin B3 on the extender. If now correct, reinspect the ports circuit. Measure the resistance between +5 and ground. Look for touching wire wrap terminals. Repeat the continuity tests.

When +5 is correct, use the logic probe to check all **H** and **L** connections in Figure D.4. Put the probe directly on IC pins. The most likely problem is omitting a connection. After fixing any problems, leave the computer open and the ports board on the extender.

D.4.3 Test Control-out

Enter the TESTOUT program in Section 3.5.1. Run the program for port 768 (control-out). Output zero and check the CO bits with the logic probe at the breadboard end of the ribbon cable. They should all be low. Whether the values are correct or not, output 255 and check all bits for high. Then, check individual bits by outputting 1, 2, 4, 8, etc. If the results are completely wrong, immediately check the control-out 74373's Q's with the logic probe. Use Figure D.4 as a guide. If correct values are coming out of the latch/buffer, the problem must be a bad ribbon cable or incorrect wiring between the 74373 and DC37 connector. Correct the errors, recheck all bits, and proceed to Section D.4.4.

If control-out basically works but a few bits are backward, fix the ribbon cable or the incorrect connections between the wire wrap terminals and the control-out 74373 or between the 74373 and DC37 connector. Then, use the program to recheck all bits and proceed to Section D.4.4.

On the other hand, if the 74373's Q's are wrong, it's necessary to thoroughly evaluate the ports circuit. Enter the following infinite loop program:

```
      CLS
      PRINT "Running"
ILOOP:OUT &H300,0
      GOTO ILOOP
      END
```

Test Port Number Decoding

Run the program. Connect oscilloscope channel 1 via a test clip to 74138 pin 1. The signal there is IOW* from the expansion slot. It should briefly go low each time around the loop and produce a stable scope display. If not, save the program, turn the power off, remove the ports circuit from the extender card, and connect expansion slot pin B14 directly to the scope. Restart the computer and program. Check IOW* again. It must briefly go low if the program repeatedly executes OUT's to any port number.

> When working with the computer open, it's essential to be organized and disciplined. Carelessness might lead to computer damage. Troubleshooting must never be rushed.

Reinstall the ports circuit and restart the computer and output program. If IOW* is still wrong, check the connection between wire wrap terminal B14 and pin 1 of the 74138. If it's correct, the problem must be a defective 74138 crashing IOW*. Remove the chip and check again. If now correct, replace the IC. There are no other reasons for an incorrect IOW*.

The strategy here is to establish one signal which definitely works and build from there.

After IOW* checks out, connect oscilloscope channel 2 to 74138 pin 4. The signal there is DO*, the output of port number decoding circuitry. As shown in Figure 3.9, it should activate a bit longer than IOW*. If correct, proceed to *Test the 74138*.

If DO* is wrong, systematically evaluate the decoding circuitry as follows. With IOW* still connected to channel 1, touch channel 2's probe via a test clip to each input of NOR-B1 and NOR-B2. Use the wire wrap drawing to locate pins. All should be low while IOW* is active. If any isn't, check the wiring. Next, look at NOR-A1. A_8 and A_9 should both be high and A_{10} and A_{11} both low. If any is wrong, check the wiring. Once these NOR inputs are correct, check NOR-A2. All its inputs should be low when IOW* is active. If not, the problem is with the outputs of NOR-A1, B1 or B2 or with INVERTERS A1, A2 or A3. Finally, check INVERTER-A6.

The goal of this procedure is to isolate something which is definitely wrong and then inspect the few possibilities for a cause. A problem with wire wrap circuits is the inability to easily undo a connection to determine whether the problem is with the signal's origination or destination(s). One way to deal with the possibility is to systematically remove destination IC's and observe the effect. (It should be safe to extract expansion board chips with the computer's power on.)

Test the 74138

Still running the infinite loop program and triggering the oscilloscope with IOW* connected to channel 1, use channel 2 to check IOR* (at 74138 pin 2) for high. Then, verify that A_0 is low at 74138 pin 3. If these signals are correct and the chip is not defective, output Y_2 must equal IOW*. Finally, verify that the control-out 74373's enable is the inverse of Y_2 (as shown in Figure 3.9). If everything checks out, the only possible cause of a problem is the data connections. Use the scope to look at each data bit as it arrives at the control-out 74373. For the current program, all bits should be low while IOW* is active. If not, check the connections between the wire wrap terminals and 74373. A possible problem is that the data lines are wired to the 74373 outputs, not the inputs. Next, check the 74373's outputs for all lows. Finally, look at the other end of the ribbon cable for all lows. When everything checks out, change the infinite loop program to output a different value and again verify the control-out bits.

In Microsoft QuickBASIC, an infinite loop is exited by typing CTRL-BREAK.

After all errors are corrected, load the original TESTOUT program from Section 3.5.1. This time it should work. If not, start the entire troubleshooting sequence over (after a rest).

D.4.4 Test Data-out

The only difference between control-out to port 768 and data-out to port 769 is the value of A_0 which is high in the latter case. Run the test program for port 769. Since control-out works, the likelihood is that data-out will also. If not, enter an infinite loop program which outputs zero to port &H301. Run the program. Put IOW* on scope channel 1 and A_0 at 74138 pin 3 on channel 2. The latter should be high. Then, check Y_6 and the data-out 74373's enable. Problems should be easy to isolate.

D.4.5 Test Control-in

Enter the TESTIN program in Section 3.5.2. Wire CI_0 low on the applications end of the ribbon cable. Leave the other bits disconnected. Run the program for port 768, control-in. Since disconnected inputs are seen as high, the input value should be 11111110 or 254. Whether the correct value is obtained or not, disconnect CI_0 and wire CI_1 low. Input the value. Repeat for CI_2 and so on. Expected results are in a table in Section 3.5.2. If only a few bits are wrong, check the connections between the wire wrap terminals and the control-in 74373 and between the 74373 and DC37 connector. Check the ribbon cable. The problem should be easy to isolate and fix. Then, go to Section D.4.6.

If, on the other hand, no or few bits are correct, use the logic probe on the control-in 74373's D inputs to be sure the correct value is reaching the chip from the breadboard. If not, make the obvious corrections. Then recheck each bit and proceed to Section D.4.6.

If the bits reaching the 74373's D's are correct but the input value is wrong, it's necessary to analyze the ports circuit. Enter the following infinite loop program.

```
      CLS
      PRINT "Running"
ILOOP: X = INP(&H300)
      GOTO ILOOP
      END
```

Check Port Number Decoding

Run the program. Connect IOR* as it reaches 74138 pin 2 to oscilloscope channel 1. It should briefly go low once each time around the loop and produce a stable scope display. If not, check the connection from the wire wrap terminal to the 74138. When IOR* is correct, connect DO* at 74138 pin 4 to channel 2. It should go active with IOR* as shown in Figure 3.8.

Check the 74138

Still running the input program with the oscilloscope triggering on IOR*, use channel 2 to verify that IOW* is inactive and A_0 is low. If so, 74138 output Y_1 must follow IOR*. Verify this by connecting scope channel 2 to the control-in 74373's OC* at pin 1. If incorrect, check the connection between the 74138 and 74373. Because only one device at a time can put values on $d_7...d_0$, look at data-in's OC* and see if it stays high. Since control-out works, it's likely that control-in will also except for straightforward wiring errors or defective IC's.

Once these tests are done, go back to the TESTIN program and see if correct values are obtained when individual bits are wired low. The only possible remaining problem is CI connections between the breadboard and DC37 connector, or between the connector and control-in 74373, or between the 74373 and wire wrap terminals.

D.4.6 Test Data-in

Run the original TESTIN program for data-in port 769 with DI_0 wired low and the other DI's disconnected. If the expected 254 is found, wire the other bits low one at a time and check for correct input values. If they are all correct, testing and troubleshooting is complete.

On the other hand, if incorrect values are found, check the DI connections between the breadboard and DC37 connector, between the connector and data-in 74373, and between the 74373 and wire wrap terminals.

If no problem is found, modify the infinite loop program above to input from &H301. Run the program and check the circuitry which produces data-in's OC*. Also, be sure control-in's OC* is not active. These steps should quickly reveal any final problems.

Save the various test programs to analyze possible future problems with the circuit and in case another board is constructed.

D.4.7 Other Tests

After the ports work, it's a good idea to carry out the various exercises suggested in Sections 3.5.3, 3.5.4 and 3.5.6. It's especially important to measure how fast the computer and software generate control signals as discussed in Section 3.5.3.

Appendix E   Apple II Parallel Ports

Section 3.3 explains the design of a pair of 8-bit input/output ports for the Apple II family of computers. Figure 3.19 is the complete circuit. This appendix is a guide to building the ports. Section E.1 lists components and presents a layout. Section E.2 suggests an interfacing cable and connectors. Section E.3 gives wiring instructions. And Section E.4 outlines testing and troubleshooting. The objective is a complete presentation of one construction approach. Details such as connectors and cables are somewhat arbitrary and other schemes may be substituted.

E.1 Components and Layout

The circuit in Figure 3.19 requires the following components.
(1)   A Vector Model 4609 wire wrap board for the Apple II expansion slot.
(2)   Four 74LS373 latch/buffers.
(3)   One 74LS04 INVERTER.
(4)   One 74LS138 decoder.
(5)   Six IC sockets: four 20-pen, one 16-pin and one 14-pin.
(6)   A bag of Vector T44 wire wrap terminals.
(7)   Three .01 μf disc capacitors.
(8)   A wire wrap tool and an assortment of wire.
Vendors for these components are in Appendix B.

Figure E.1 suggests a layout for the circuit on the expansion board. The view is from the components side with the rear of the computer on the right. Along the bottom, just above the plug, are pads for wire wrap terminals. When inserted, the terminals allow expansion slot signals to be connected to the ports circuit. The lower row starts with 1 on the left and ends on the right at 25 (which is +5). The upper row starts on the right with 26 (which is ground) and ends on the left at 50.

The following steps prepare the board for wiring.
(1)   Solder wire wrap terminals to the following expansion plug pads: 2, 3, 18 and 41-49. Dots in Figure E.1 show the approximate locations. A terminal is pushed through from the components to the solder side.
(2)   Epoxy the six wire wrap sockets approximately as shown in the layout.

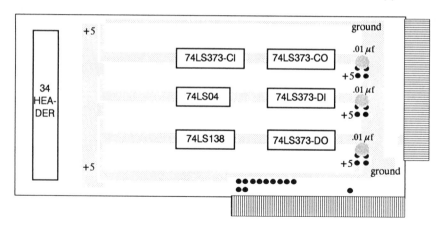

Figure E.1. Layout for the Apple II I/O ports circuit. The view is from the components side and the rear of the computer is on the right.

(3)   Finally, place three pairs of wire wrap terminals between the right vertical ground strip and +5 pads as shown in Figure E.1. Then, in pads just above the terminals, solder .01 μf capacitors between +5 and ground (for power supply noise reduction). +5 and ground originate from the computer via the expansion plug.

E.2 Connectors and Cable

It's necessary to construct a cable to run from the ports (in the computer) to various applications on breadboards and wire wrap cards. Recall from Chapter 3 that the two 8-bit I/O ports present 32 connections. Since ground is also required, a 34-conductor ribbon cable is selected.

> *A common practice is to use multiple gound lines, sometimes every other wire in circuits with high frequency signals. Extra grounds have proven unnecessary here even for 20-foot cables.*

A 34-conductor edge socket is chosen for both ends of the ribbon cable. The expansion board receives the socket with a 34-conductor header. The breadboard or wire wrap card on the applications end receives the socket with an identical header. Pin assignments for the headers, henceforth called the parallel ports connector, are in Figure E.2. (When used on a breadboard, it's necessary to slightly bend the header's pins.)

> *Vector's expansion board contains a 40-pin card-edge plug to which the ribbon cable might be connected. However, when inserted in most model Apple II's, the plug is inaccessible.*

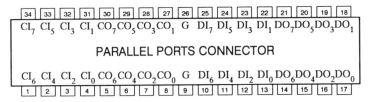

Figure E.2. Pin assignments for the parallel ports connector.

The specific connector and cable components are:

(1) Some length of 34-conductor ribbon cable.

(2) Two 34-conductor .100" by .100", flat cable, edge sockets (such as 3M series 3414).

(3) Several .100" by .100" 34-pin headers, wire wrap type, for both the expansion board and applications circuits (such as 3M series 3594).

Vendors are listed in Appendix B.

Mount the connectors as follows:

(1) Epoxy a header on the left side of the expansion board approximately as shown in Figure E.1. Pin 1 is at the top left.

(2) Mount sockets on both ends of the ribbon cable. Connectors designed for flat cables are mounted with a vise. Just observe what has to happen and use common sense.

A good idea is to run the cable between the expansion board and a breadboard and employ a resistance measuring multimeter to test continuity. Use the pin assignments in Figure E.2.

E.3 Wiring

The next step is to connect the IC sockets and terminals. Two "wire wrap" drawings, the first in Figure E.3, facilitate the work. They show the header's 34 pins on the left, the expansion plug's 12 wire wrap terminals, and the 6 IC sockets, all from the components side. The arrangement is similar to the layout in Figure E.1 except the +5 and ground terminals are omitted.

In both wire wrap drawings, bold characters are placed above and below pins and terminals. Points to be connected are labeled by the same character.

Another approach is to use drawings similar to Figure E.3 but "up-side-down-and-backward." This is because the wire wrap card has to be turned over to make connections. The "right-side-up" method here encourages thinking about each step in terms of what's being wired. And it facilitates testing and troubleshooting as described in Section E.4.

E.3.1 Wire +5 and Ground

Use Figure E.3 to wire +5 to each **H** labeled pin and ground to each **L** pin. Start on the right with the separate +5 and ground wire wrap terminals (not shown in the drawing). Wire across a row of IC's. It's important not to make ground "loops." Included in the **H**'s and **L**'s are IC pins which must be wired high or low as indicated in Figure 3.19. Also, don't forget the **L**'s on the header. It's customary to use red wire for +5 and black for ground.

E.3.2 Wire $d_7...d_0$

Connect the eight data lines from the expansion plug terminals to all four 74LS373 IC's. Specifically, wire all pins labeled **0** in Figure E.3. Wire all **1**'s, all **2**'s and so on up to all **7**'s. Note that $d_7...d_0$ is connected to the Q's of the IN 74LS373's and to the D's of the OUT 74LS373's.

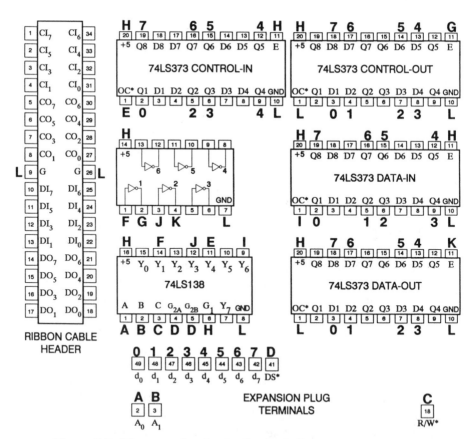

Figure E.3. Wire wrap drawing for the Apple II ports circuit showing +5 and ground connections, $d_7...d_0$, and 74138 wiring.

E.3.3 Wire Control Circuitry

Refer to Figures 3.19 and E.3 and make the following connections.

(1)   Wire address line A_0, labeled **A**, from its wire wrap terminal to 74LS138 select A.

(2)   Wire address line A_1 (**B**) from its wire wrap terminal to 74LS138 select B.

(3)   Wire R/W* (**C**) from its wire wrap terminal to 74LS138 select C.

(4)   Wire DS* (**D**) from its wire wrap terminal to 74LS138 G_{2A}* and G_{2B}*.

(5)   Wire 74LS138 output Y_4 (**E**) to the control-in 74LS373's OC*.

(6)   Wire 74LS138 output Y_1 (**F**) to INVERTER-1. Wire INVERTER-1's output **G** to the control-out 74LS373's E.

(7)   Wire 74LS138 output Y_6 (**I**) to the data-in 74LS373's OC*.

(8)   Wire 74LS138 output Y_3 (**J**) to INVERTER-2. Wire INVERTER-2's output **K** to the data-out 74LS373's E.

E.3.4 Wire the Cable Header

Finally, use Figure E.4 to wire the 32 lines from the cable header to the respective 74LS373's. Specifically:

(1)   Wire pins labeled **0** to **7** to the control-in 74LS373's data inputs.

(2)   Wire pins labeled **A** to **H** to the control-out 74LS373's Q outputs.

(3)   Wire pins labeled **I** to **P** to the data-in 74LS373's D inputs.

(4)   Wire pins labeled **Q** to **X** to the data-out 74LS373's Q outputs.

This completes construction. Visually inspect the work. Install the IC's.

E.4 Testing and Troubleshooting

The following tools are needed to test and troubleshoot the ports circuit: a logic probe, IC test clips (14-, 16- and 20-pin), a multimeter (which measurers voltage and resistance), an oscilloscope, and an extender card for the Apple II expansion slot.

As discussed in Section 3.3, the control and data-in ports respond to peeks from specific addresses and the control and data-out ports respond to pokes to different addresses. As explained in Chapter 3, if the board is in slot 3, the addresses are: CI = 49328, CO = 49329, DI = 49330 and DO = 49331.

E.4.1 Continuity Tests

Before installing the circuit, test all wiring by verifying with the multimeter that the resistance between points which should be connected is zero. Use Figures 3.19, E.3 and E.4. Follow the order used in construction. Put multimeter probes directly on IC pins to detect the possibility of a defective socket or bent pin.

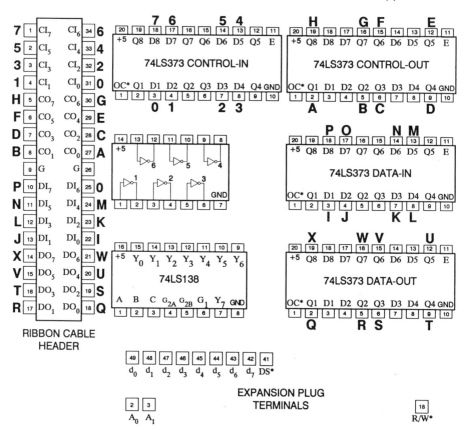

Figure E.4. Wire wrap drawing showing connections between the ribbon cable header and the four 74373's.

If not carried out earlier, run the ribbon cable from the I/O ports to a breadboard. Then, use Figures E.2 and E.3 to test the continuity of all connections.

E.4.2 Test +5 and Ground

The next step is to check +5 and ground. Before doing so, read the computer's manual on installing expansion slot cards. Always follow instructions one of which is to have the power off when inserting or removing a card.

Open the computer and insert the extender card. Next, install the ports in the extender. Turn the power on. If any chip rapidly heats up, turn the power off. If the computer doesn't boot properly, turn the power off. In either case, remove the ports circuit and repeat the continuity tests. Look for bent wire wrap pins or a solder bridge. If no problems are found, remove the

IC's and cable, reinsert the board and turn the power on. The computer should boot. Next, reinstall the IC's one-by-one rebooting each time. When the computer doesn't boot or the just installed IC rapidly heats up, the chip is either defective or incorrectly wired.

Once the computer boots with the ports installed and no problems are apparent, use the multimeter to measure +5 at some point on the expansion board. If it isn't at least 4 volts, turn the computer off and remove the board. Turn the power back on and measure +5 at pin 25 on the extender card. If now correct, reinspect the ports circuit. Measure the resistance between +5 and ground. Look for touching wire wrap terminals. Repeat the continuity tests.

When +5 is correct, use the logic probe to check all **H** and **L** connections in Figure E.3. Put the probe directly on IC pins. The most likely problem is omitting a connection. After correcting any errors, leave the computer open and the ports board on the extender.

E.4.3 Test Control-out

Enter the following Applesoft version of the TESTOUT program in Section 3.5.1.

```
10  REM      Program TESTOUT
20  CI = 49328
30  CO = CI+1 : DI = CI+2 : DO = CI+3
35  HOME
40  INPUT "Enter CO or DO "; OP$
50  IF OP$ = "CO" THEN AD = CO
55  IF OP$ = "DO" THEN AD = DO
60  INPUT "Enter 0 to 255 or > 255 to End ";OV
70  IF OV > 255 GOTO 90
80  POKE AD,OV
85  GOTO 60
90  END
```

Run the program. Enter CO and then 0. Check the eight control-out bits with the logic probe at the breadboard end of the ribbon cable. They should all be low. Whether the values are correct or not, output 255 and check all bits for high. Then, check individual bits by outputting 1, 2, 4, 8, etc. If the results are completely wrong, immediately check the control-out 74373's Q's with the logic probe. Use Figure E.4 as a guide. If the values are correct, the problem must be incorrect wiring between the 74373 and cable header or a bad ribbon cable. Correct errors, recheck all bits and proceed to Section E.4.4.

If control-out basically works but a few bits are backward, fix the bad connections. Then, use the program to recheck all bits and proceed to Section E.4.4.

On the other hand, if the 74373 Q's are wrong, it's necessary to thoroughly evaluate the ports circuit. Enter the following infinite loop program:

```
10   HOME
15   CO = 49329
20   PRINT "Running"
30   POKE CO,0
40   GOTO 30
50   END
```

Run the program. Connect oscilloscope channel 1 via a test clip to the 74138's pin 4. The signal there is DS* from the expansion slot. It should briefly go low each time around the software loop and produce a stable scope display. If not, save the program, turn the power off, remove the ports circuit from the extender card and connect expansion slot pin 41 directly to the scope. Restart the computer and program. Check DS* again. It must briefly go low if the program repeatedly executes POKE's to any address in the expansion slot's range. If necessary, exit the program and check the software address.

Applesoft infinite loops may be exited by typing CTRL-C. When working with the computer open, it's essential to be organized and disciplined. Carelessness might lead to computer damage. Troubleshooting must never be rushed.

Reinstall the ports circuit and restart the computer and infinite loop program. If DS* is still wrong, check the connections between wire wrap terminal 41 and pins 4 and 5 of the 74138. If they are correct, the problem must be a defective 74138 crashing DS*. Remove the chip and check again. If now correct, replace the IC. There are no other reasons for an incorrect DS*.

The strategy here is to establish one signal which definitely works and build from there.

After DS* checks out, connect oscilloscope channel 2 to 74138 pin 3. The signal there is R/W* from the expansion slot. As shown in Figure 3.17, it should be active while DS* is active. If R/W* is incorrect, check the wiring and 74138. Not much else could be wrong.

Still running the infinite loop program and triggering the oscilloscope on DS* connected to channel 1, use channel 2 to check A_0 at 74138 pin 1. It should be high while DS* is active. Then, verify that A_1 is low at 74138 pin 3. If these signals are correct and G_1 is wired high, unless the 74138 is defective, output Y_1 must be correct. Verify at pin 13 that the signal is the same as DS*. Finally, verify that the control-out 74373's enable is the inverse of Y_1 (as shown in Figure 3.17). If everything checks out, the only possible cause of a problem is the data connections. Use the scope to look at each data bit as it arrives at the control-out 74373. For the current program, all bits should be low while DS* is active. If not, check the connections between the wire wrap terminals and 74373. A possible problem is that the data lines are wired to the 74373 outputs, not the inputs. Next, check the 74373 outputs for all low. Finally, look at the other end of the ribbon cable for all lows. When everything checks out, change the infinite loop program to output a different value and again verify the control-out bits.

The goal of this procedure is to isolate something which is definitely wrong and then inspect the few possibilities for a cause. A problem with wire wrap circuits is the inability to easily undo a connection to determine whether the problem is with the signal's origination or destination(s). One way to deal with the possibility is to systematically remove destination IC's and observe the effect. (It should be safe to extract expansion board chips with the computer's power on.)

After all errors are corrected, load the TESTOUT program. This time it should work. If not, start the entire troubleshooting sequence over (after a rest).

E.4.4 Test Data-out

The only difference between control-out to address 49329 (C0B1 hex) and data-out to address 49331 (C0B3) is the value of A_1 which is high in latter case. Run the test program for DO. Since control-out works, the likelihood is that data-out will also. If not, load the infinite loop program and change the poke to address 49331. Run the program. Put DS* on scope channel 1 and A_1 on channel 2. The latter should be high. Then, check Y_3 and the data-out 74373's enable. Problems should be easy to isolate.

E.4.5 Test Control-in

Enter the following Applesoft version of the TESTIN program in Section 3.5.2.

```
10   REM     Program TESTIN
20   CI = 49328
30   CO = CI+1 : DI = CI+2 : DO = CI+3
35   HOME
40   INPUT "Enter CI or DI "; OP$
50   IF OP$ = "CI" THEN AD = CI
55   IF OP$ = "DI" THEN AD = DI
60   IV = PEEK(AD)
65   PRINT IV
70   INPUT "Any Key to Repeat, END to Stop ";A$
75   IF A$ = "END" GOTO 90
```

```
80   GOTO 60
90   END
```

Wire CI_0 low on the applications end of the ribbon cable. Leave the other bits disconnected. Run the program. Since disconnected inputs are seen as high, the input value should be 11111110 or 254. Whether or not the correct value is obtained, disconnect CI_0 and wire CI_1 low. Input the value. Repeat for CI_2 and so on. Expected results are in a table in Section 3.5.2. If only a few bits are wrong, check the connections between the expansion board's cable header and the control-in 74373 and between the 74373 and wire wrap terminals. If necessary, check the ribbon cable. The problem should be easy to isolate and fix. Then, go to section E.4.6.

If, on the other hand, no or few bits are correct, use the logic probe on the control-in 74373's D inputs to be sure the correct value reaches the chip from the breadboard. If not, make the obvious corrections. Then, recheck each bit and proceed to Section E.4.6.

If the bits reaching the 74373's D's are correct but the input value is wrong, it's necessary to analyze the ports circuit. Enter the following infinite loop program.

```
10   HOME
15   CI = 49328
20   PRINT "Running"
30   IV = PEEK(CI)
40   GOTO 30
50   END
```

Run the program. Connect DS* as it reaches 74138 pin 4 to oscilloscope channel 1. Connect 74138 output Y_4 to channel 2. It should equal DS*. Check the signal again at the control-in 74373's OC*. Because only one device at a time can put values on $d_7...d_0$, look at data-in's OC* and see if it stays inactive. Since control-out works, it's likely that control-in will also except for straightforward wiring errors or defective IC's.

Once these tests are done, go back to the TESTIN program and see if correct values are obtained when individual bits are wired low. The only possible remaining problems are the connections between the breadboard and expansion board header, between the header and control-in 74373 and between the 74373 and wire wrap terminals.

E.4.6 Test Data-in

Run the TESTIN program for DI with DI_0 wired low and the other bits disconnected. If the expected 254 is found, wire the other bits low one at a time and check for correct input values. If all are correct, testing and troubleshooting is complete.

On the other hand, if incorrect values are found, check the DI connections.

If no problem is found, modify the infinite loop program above to peek from address 49330. Run the program and check the circuitry which produces data-in's OC*. Also, be sure control-in's OC* is not active. These steps should quickly reveal any final problems.

Save the various test programs to analyze possible future problems with the circuit and in case another board is constructed.

E.4.7 Other Tests

After the ports work, it's a good idea to carry out the various exercises suggested in Sections 3.5.3, 3.5.4 and 3.5.6. It's especially important to measure how fast the computer and software generate control signals as discussed in Section 3.5.3. Also, 6502 assembly language should be used if faster operations are required.

Appendix F   Computer Architecture

To fully understand the design of the I/O ports described in Sections 3.2 and 3.3 and the software presented in Section 3.5, it's necessary to have some knowledge of microcomputer architecture. This appendix defines components and outlines operations.

F.1 Components

Figure F.1 is a block diagram of a typical microcomputer. The following sections describe the six basic components.

CPU

The Central Processing Unit or microprocessor is an integrated circuit which executes programs and in doing so controls external memory and I/O circuits. In addition to +5, ground and possibly other power supply voltages, the processor has some number of data and address pins and a variety of controls.

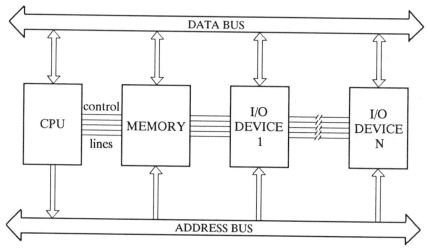

Figure F.1. Block diagram of a typical microcomputer.

For instance, the Intel 80286 has 16 data, 24 address and 18 control connections.

CPU's interact with memory and I/O devices through the data and address buses and the controls. Specifically, during program execution, when a statement (in whatever language) requires a memory read or write or an input or output, the CPU carries out an appropriate timing sequence (or machine cycle) on its control lines and buses. There are only a small number of cycles and each is specified by the CPU's manufacturer. Memory and I/O circuits are designed to respond to the specifications.

Computers with Intel microprocessors (PC's and PS/2's) have four machine cycles of interest here: memory read, memory write, input from an I/O port, and output to a port.

Computers with Motorola microprocessors (Apples and Macintoshes) have only two cycles of interest: memory read and memory write.

Data Bus

The data bus may be thought of as a set of wires connected to the microprocessor's data pins and to the memory and I/O circuits (as shown in the figure). The CPU sends values to and retrieves values from memory and it outputs values to and inputs values from I/O devices all over the data bus. The size of the values (and bus) depends on the number of microprocessor data connections: 8, 16 or 32 bits. Binary data values, the CPU's data pins and the bus wires are designated $d_7...d_0$ (in 8-bit machines), $d_{15}...d_0$ (in 16-bit computers) or $d_{31}...d_0$ (in 32-bit systems). Only 8-bit values are needed for the I/O circuits in Chapter 3. If more data lines are available, only the low eight ($d_7...d_0$) are used.

Address Bus

The address bus may be thought of as a set of wires connected to the CPU's address pins and to the memory and I/O circuits. The CPU specifies a memory location at which a data value is saved or from which it is retrieved by supplying a binary number to its address pins and therefore the bus. The maximum number of locations depends on the number of address bits. The common cases are 16, 24 and 32 which allow up to 64 thousand, 16 million and 4 billion locations, respectively. Address values, the CPU's address pins and the bus wires are designated $A_{15}...A_0$, $A_{23}...A_0$ or $A_{31}...A_0$.

Control Lines

In addition to data and address buses, the CPU originates a number of signals which control the memory and I/O circuits during the various machine cycles. The CPU also receives signals from memory and I/O circuits which are monitored for information such as wait.

Memory

Memory serves many purposes in computer operations. For instance, the machine code for all software including the operating system is located in specified ranges of random access and read only memory. Also, at various times, the memory contains items such as program text, all sorts of data, buffers which temporarily store sets of values as they pass to and from disk storage, and, in some cases, the contents of the monitor screen. One of the important CPU operations is maintaining internal registers which keep track of the addresses relevant to the current program.

Memory circuits for a particular computer are designed to respond to the read and write machine cycles generated by its CPU (as discussed below).

I/O Devices

Common input/output devices are the computer's keyboard, mouse, disk controller, game port, printer port, etc. Most data acquisition systems are also I/O devices.

For computers with Intel microprocessors, I/O circuits are designed to respond to the input and output machine cycles generated by the CPU (as discussed below). Motorola microprocessors do not have separate input/output cycles and so I/O devices are designed to act like memory and respond to read and write cycles (with addresses specified by the computer's manufacturer, as discussed below).

The six components above are sufficient for the purposes of understanding and designing I/O circuits. However, for completeness, two other operations are introduced. Most microprocessors allow an external circuit (connected to the data, address and control signals) to take over and become a Temporary Bus Master. The TBM then runs the computer by executing machine cycles identical to the CPU's. An example is a Direct Memory Access (DMA) circuit which can tell the CPU (via control lines) to get off the buses and controls, after which it stores values in or retrieves values from the memory by executing cycles which to the memory appear the same as the CPU's. Another feature is interrupts. Microprocessors allow circuits connected to the address, data and control lines to interrupt current program execution and tell the CPU to execute a different program. After the interrupt routine is finished, the original program resumes.

F.2 Memory Write and Read Machine Cycles

When the CPU needs to store a value in memory or retrieve a previously stored value, it executes a memory write or read machine cycle. The sequences go roughly as follows:

Memory Write

The CPU puts the location where a value is to be stored on the address bus. (As stated earlier, a number of internal registers keep track of the addresses of program code, variables and other data, etc.) Next, the CPU puts the value on the data bus. (Depending on the details of the program, the CPU has previously obtained the value in one of a great variety of ways.) Then, the CPU, using its master clock, takes several control lines through a timing sequence which causes the memory circuit to store the value on the data bus at the location on the address bus.

Memory Read

The CPU puts the location of a previously stored value on the address bus and then activates several controls. The memory circuit responds by putting the value at the address on the data lines. The CPU then stores the value in an internal register (for later use as determined by the program).

F.3 I/O on Computers with Motorola Microprocessors

As previously stated, Motorola microprocessors do not have separate input and output machine cycles. Therefore, I/O devices occupy part of the regular memory space and respond to read and write cycles. The addresses of possible devices are specified by computer manufacturers. One case is illustrated below.

Suppose an I/O device is located in an Apple II's #4 expansion slot. The machine's designers allocated addresses 49344 to 49359 to the slot and therefore to the device. When the BASIC statement POKE address,value is executed, the CPU carries out a memory write machine cycle the first steps of which are to put address on the address bus and value on the data bus. Next, the CPU takes several controls through a timing sequence which causes the device to save value (in a way appropriate for its function). It's assumed address is in the range for slot #4 and the I/O circuit is wired to respond to address. (Devices do not have to support to all address in a slot's range. And a POKE to an unsupported address results in no action.) So when the Apple executes POKE 49344,29, the microprocessor puts the binary value of 49344 on the address lines and the binary of 29 on the data lines. In response to the controls, the I/O circuit in slot #4 stores LLLHHHLH.

When the BASIC statement variable = PEEK(address) is executed, the CPU carries out a memory read cycle the first steps of which are to put address on the address bus and to activate controls. If address is in the range for slot #4 and if the device in slot #4 is wired to respond to address, then the circuit uses the controls to put a value on the data bus. The CPU reads the value and assigns it to variable. For example, when the

Apple executes X = PEEK(49359), the CPU puts 49359 on the address lines and activates controls. The device in slot 4 puts a value on the data lines which the CPU reads and assigns to the BASIC variable X. Specifically, if the device upon decoding 49359 and a PEEK command puts LHLLLHHH on the lines, X = 71 after the operation. A PEEK of an address in a slot's range which the occupying I/O circuit does not support results in an arbitrary value assigned to X.

F.4 I/O on Computers with Intel Microprocessors

Computers with Intel microprocessors have input and output machine cycles. Just as for memory, incoming and outgoing values are put on the data bus. The question is how to specify which of many I/O devices responds to a cycle? The answer is that the CPU puts a port number on a subset of the address lines. Each I/O device monitors the addresses and, when its preassigned port number appears, acts upon the current cycle. The next question is how memory and input/output cycles are distinguished? The answer is that the CPU supplies a control signal (say, M-I/O) which is in one state during memory cycles and the other during input/output cycles. Memory and I/O circuits monitor the line and respond appropriately. The following example illustrates a specific case.

For IBM PC/XT/AT's and compatibles, port numbers are 12 bits. When the BASIC statement OUT port number,value is executed, the CPU generates an output machine cycle the first steps of which are to put port number on the lower 12 bits of the address bus and value on the data bus. Next, the CPU takes several control lines including M-I/O through a timing sequence. All I/O circuits follow the sequence. When M-I/O indicates an output cycle, the device which finds its port number on the address lines uses the other controls to save the value on the data lines (in a way appropriate for the device's function). For example, execution of OUT &H300,0 causes whatever device is wired for port 300 Hex to save 00000000 binary. In IBM PC's and compatibles, port numbers are designated for the printer port, keyboard, disk controller, etc. However, ports 300 to 31F Hex are set aside for prototype boards (and are used in Chapter 3). Outputs to non-existent ports cause no action.

When the BASIC statement variable = INP(port number) is executed, the CPU generates an input cycle the first steps of which are to put port number on the address lines, indicate input/output on M-I/O and activate other controls. All I/O circuits monitor the addresses and controls. When M-I/O indicates an input cycle, the one port which finds its preassigned number on the address bus puts a value on the data bus. The CPU reads the value and assigns it to variable. For example, in executing X = INP(&H301),

the CPU puts 301 Hex on the address lines and activates the controls for an input machine cycle. The one I/O device wired to respond to 301 Hex puts a value on the data lines, say, HLLLLLLH. The CPU reads the data lines and assigns 129 to X. If no port 301 is present, the value read is arbitrary.

F.5 Computer Expansion Slots

The machine cycles outlined above are presented in terms of the microprocessors in popular computers. The I/O ports in Chapter 3 are designed for the IBM and Apple II expansion slots. The slots make available the CPU's data and address lines and the cycles follow the outlines above. However, the expansion slot's control lines are not necessarily identical to the CPU's. Section 3.2 presents in detail the sequence of expansion slot control, data and address signals during the input and output machine cycles of whatever CPU is running an IBM PC/XT/AT or compatible (including machines with 80386 and 80486 processors which support Industry Standard Architecture or Expanded ISA). Section 3.3 presents in detail the sequence of Apple II expansion slot control, data and address signals during memory read and write cycles to addresses in each slot's allocated range.

Index

71969